GEOLOGY

About the Author

Dr Rajeeva Guhey is Professor of Geology, at Government Nagarjuna P.G. College, affiliated to Pt. Ravishankar Shukla University, Raipur, C.G. He did M. Tech. in Applied Geology from Government Engineering College, Ph.D. on Proterozoic Carbonates of Chhattisgarh Basin, from Pt. R.S.U. Raipur, in 1984 and 1993 respectively. He has completed 5 research projects on Proterozoic sedimentary carbonates from Proterozoic Chhattisgarh, Indravati and Singhora Basins, sponsored by MPCOST, UGC, DST and CGCOST. He is a recipient of **Young Scientists Award** for best paper presentation in 7" Young Scientist's Congress, at Devi Ahilya University Indore, M.P. during 1992. He has published 20 research papers in National and International Journals. He organized a National Seminar on Recent Advances in Proterozoic Basins of India during 1999, He visited Florence, Italy in 2004 to participate in **32nd International Geological Congress (IGC)**. He is a life member of South Asian Association of Economic Geologists. Geological Society of India, Bangalore, Indian Association of Sedimentologists, Aligarh. He guided number of Ph.D. students for pursuing their research work. He edited a book entitled **"Recent Researches on Proterozoic Basins of India"** and a Multiple Science Journal entitled **Research Fronts** from Govt Digvijay College, Rajnandgaon, Chhattisgarh, Presently he is teaching to undergraduate and post graduate students.

GEOLOGY

PRINCIPLES AND PRACTICAL MANUAL

Rajeeva Guhey
Professor, Geology
Govt. N.P.G. College of Science
Raipur (C.G.)

NEW INDIA PUBLISHING AGENCY
New Delhi – 110 034

NEW INDIA PUBLISHING AGENCY
101, Vikas Surya Plaza, CU Block, LSC Market
Pitam Pura, New Delhi 110 034, India
Phone: + 91 (11)27 34 17 17 Fax: + 91(11) 27 34 16 16
Email: info@nipabooks.com
Web: www.nipabooks.com

Feedback at feedbacks@nipabooks.com

ISBN: 978-93-85516-81-8

Composed, Designed & Printed in India

Dedicated to
My Loving Parents

Niteesh Kumar Dutta
Director General (Retd.)
Geological Survey of India
Hyderabad, Telangana
E-mail : niteeshdutta@gmail.com

9, Utkarsha Nagar, Katol Road,
Nagpur-440013 (M.S.)

Foreword

While I was trying to write the 'Foreword' for this "Geology : Principles and Practical Manual" by Professor Rajeeva Guhey, I was distinctly reminded of my initial days in the University Teaching Department (U.T.D) , University of Saugar (presently Dr. Hari Singh Gour Sagar Central University, Sagar) for the Bachelor of Science degree with Geology as one of the subjects of science. In those days, there was nothing like a subject called Geology or Earth Science, though some Geography *(Bhoogol)* was inducted into us in the subject course called 'Social Studies' *(Samajik Adhyayan)* with emphasis on the political boundaries and a bit of physiography of the various districts and states, ultimately of India and World as a whole.

The subject of Geology was totally new to us. We were from Hindi medium school and were finding it difficult to cope up with this new subject, which in simple terms is **'Geo'** meaning 'Earth', and **logos'** meaning 'science'. Thus, struggle for survival began, that is to score first division marks in the science subjects. (The other subjects were more familiar subjects of physics and chemistry). We had to search for the various topics of Geology like 'Solar System', 'Origin of Solar System and Earth', 'Permanency of Continents and Ocean Basins', and somewhat totally new topic 'Isostasy' etc. in whatever book which were coming handy on us, those days. I remember that one book 'Physical Geology' by Professor A.K. Dutta of Bokaro College was very useful with easy to understand language and containing all so called 'topics of related importance'. The book on 'Physical Geology' by Dr. Arthur Holmes was also recommended but it was too exhaustive and difficult (in language) for the no vice students. We were always striving hard in preparation of 'notes' on various important topics for scoring 'those times impossible' 60% marks in Geology particularly. There was no collection of topic wise explanatory notes for us to go through.

It is in this perspective, this "Geology : Principles and Practical Manual" by Professor Rajeeva Guhey comes in very handy to the aspiring students. It is a compendium of all the required information for these students on various important subjects of this vast subject of 'Geology'. The 'present is clue to the past' is well established law in our subject of natural science. The well established 'hypotheses' and theories' which define the Earth's phenomenon, are applicable to the other planets with some variation based on the atmospheric conditions and their distance from the Sun. Hence, good grounding and grasp in the raw minds of the new students are essential for them to understand the various phenomenon which will make their basic understanding of the subject very strong.

Dr. Rajeeva Guhey had been an outstanding student of Government Science College, Raipur which culminated with a Doctor of Philosophy degree on the Carbonate rocks of the Chhattisgarh region itself, where he had his upbringing and grooming. He has teaching experience of undergraduate & post-graduate classes. He has groomed so many young geologists by his teaching in various colleges of Chhattisgarh.

This "Geology : Principles and Practical Manual" is an outcome of his vast teaching experience, which is so much apparent in its content and presentation. In this, Dr. Guhey has tried to incorporate all the necessary components of the subject, for which he deserves all accolades. There is still the scope of incorporation of more modern topics of plate tectonics etc., which I know is in the mind of Prof. Guhey.

I wish that his efforts are highly useful to the budding geologists of the region and to the whole community of Earth Scientists and Mining Engineers, ultimately.

Dr.Rajeeva Guhey should feel highly satisfied with his unique attempt to drive on a difficult tract of writing a comprehensive compendium on this vast and variegated, extremely interesting subject called 'Geology' for whose cause, he has devoted himself fully till date.

Niteesh

Niteesh Kumar Dutta

Nagpur, Maharashtra
Date: 27-12-2016

Preface

During my long teaching experience, I felt that geology is a new subject for beginners. The undergraduate students of this stream are deprived of concise and comprehensive text book particularly on principles and practical aspects of geology. They have no other option than to go through numerous text books on different branches of geology and are often confused as to how much they should read and write, it leads to wastage of time and energy.

With this objective in view, I have written this book entitled "Geology: Principles and Practical Manual". An attempt has been made to discuss the fundamental principles of different branches of geology prescribed in the syllabus, so that the students acquire basic knowledge of the subject. The book contains 8 chapters, viz. Geomorphology, Geological Maps, Crystallography, Mineralogy & Petrology, Palaeontology, Stratigraphy, Structural Geology and Economic Geology, Remote Sensing & Photogeology. Each chapter consists of specific branch of geology and deals with the principles and then its practical aspects. Chapter-1, Geomorphology deals with definition and introduction of different natural processes of landforms, and then practical exercises on different geomorphological models with labelled figures and precise descriptions. Chapter-2, Geological Maps deals with different terminologies used in geological map like contour line, outcrop, dip, strike etc. Different practical exercises with some solutions on geological maps: three point problem and cross section map, so that students can learn how the maps are prepared and interpreted. Chapter-3, Crystallography contains basic terminologies, symmetry elements and different crystal systems etc. which will be helpful in identification of crystals. Crystals of important minerals with their faces and Miller's symbols and figures are also incorporated in this chapter. Minerals and Rocks are dealt in Chapter-4, it comprises physical and optical properties of important rock forming minerals with coloured photos and photomicrographs. In petrology section, characteristic properties of all the rock types (Igneous, Sedimentary and Metamorphic) have been described briefly in tabular form with coloured diagrams; it would be of great help to the students to identify different rocks during practical exercises. Chapter-5, Palaeontology contains brief description on modes of fossil preservation and description of fossils from different phylum including plant fossils, their classification morphology and geological age with diagrams. Chapter-6, Stratigraphy covers principles, classification and stratigraphic correlation. It also contains Physiographic divisions of India, Geological Time Scale, and Indian stratigraphy from Archaeans to Dharwar to the Siwalik Formation with their geological successions and distribution in the map of India. Chapter-7, Structural Geology deals with identification and description of structural features including the primary sedimentary features like bedding, cross bedding etc. their attitude (strike and dip), unconformity, joints and the diastrophic structures like fold and fault with block diagrams. The last Chapter-8, Economic and Ore Geology, Remote sensing and Photogeology consists of basic concepts and practical aspects of these subjects as prescribed syllabus of B.Sc./ B.E. Civil and Mining Engineering courses in various Universities and Institutes.

Since this is my first effort, there ought to be several mistakes and shortcomings. The critical evaluation and suggestions by the learned readers will be wholeheartedly welcome, which will go towards the improvement of this book .

I shall remain indebted to Shri Niteesh Kumar Dutta, Director General (Retd.), Geological Survey of India for his constant encouragement and valuable suggestions. Thanks are also to late Dr. Ashok Kumar Pandey, Principal, Government N.P.G. College of Science, Dr S.S. Bhadauria, Head of Department of Geology, Prof. Sridhar Rao, Prof. Pradeep Thakur and Dr. Sandeep Vansutre, Department of Geology, , Dr Reeta Soni, Professor English Govt. N.P.G. College of Science, Raipur for their inspiration and constant guidance.

I owe my special gratitude to my Professors late Dr W.N. Bansode, Dr G.S. Parashar, late Dr T.P. Paikra, and my Ph.D. supervisor late Dr N.P. Wadhwa, Professor and Head, Govt. Engineering College, Raipur, for inspiring me for my academic pursuit. I am thankful to Dr Rajendra Singh, Principal, Govt. College Sarangarh, C.G. and Prof. Pradeep Singh Gaur, Head, Deptt of Geology, Govt. College, Kanker, C.G. for their unconditional support to complete this task.

I have no words to express my sincere feelings and thankfulness to my wife Dr. Arti Guhey, and sons Abhishek and Anilabh, for their constant accompaniment and providing an atmosphere for creativity at home to complete this huge task.

Dr Rajeeva Guhey
Professor, Geology
Govt. N.P.G. College of Science
Raipur, Chhattisgarh

Contents

Chapter 1

Geomorphology

1.1 Geomorphology

Geomorphology is the scientific study of the origin and evolution of topographic and bathymetric features created by physical or chemical processes operating at or near the earth's surface.

The surface of the earth is unstable and affected by a combination of surface and geologic processes that cause tectonic uplift and subsidence. **Surface processes** are mainly physical and chemical action of water, wind, ice, igneous activity and activities of living organisms on the surface of the earth. Upliftment of mountain ranges, the growth of volcanoes, isostatic changes in surface elevation and formation of deep sedimentary basins are categorized under **Geological processes**.

Main geomorphological processes occurring in nature are as follows:

1.1.1 Aeolian processes

Aeolian processes include activities of the winds, its ability to shape the surface of the earth. Wind acts as an effective agent of erosion, transportation and deposition materials, and large supply of fine, unconsolidated sediments. Aeolian processes are important in desert (arid) environments.

1.1.2 Fluvial processes

Moving water in a river channel, is able to mobilize sediment and transport it downstream. The rate of sediment transport depends on the availability of sediment itself and the discharge by river.

River originates from mountain with V shaped valley, which later increases in size, merging with other rivers. The network of rivers thus formed is known as drainage system. These systems show four general patterns viz. dendritic, radial, rectangular, and trellis etc. In the mature stage, the river form alluvial flood plains, meanders and ox-bow lakes. Finally when it merges to ocean, forms delta.

1.1.3 Glacial processes

Glaciers, while geographically restricted, are mainly moving ice and effective agents of landscape formation and change. The gradual movement of ice down a valley causes abrasion and plucking of the underlying rock. Abrasion produces fine sediment, termed glacial flour. The debris transported by the glacier are exposed when the glacier recedes, are known as moraine. Glacial erosion is responsible for U-shaped valleys, as opposed to the V-shaped valleys of fluvial origin.

1.1.4 Igneous processes

Plutonic and volcanic activities have important impact on geomorphology of an area. Plutonic rocks intruded and consolidated at depth cause upliftment and subsidence of the earth's surface.

The action of volcanoes tends to rejuvenate landscapes by covering the old land surface with lava and tephra, releasing pyroclastic material and some times forcing the rivers through new paths. The cones built up by volcanic eruptions also bringing in substantial changes in topography.

1.1.5 Biological processes

Living organisms are also agents of the geomorphic processes, ranging from biogeochemical processes controlling chemical weathering, to influence the mechanical processes by burrowing and tree roots on soil development. It even controls the global erosion rates through carbon dioxide balancing.

1.1.6 Tectonic processes

The effects of tectonics on landscape are mainly dependent on the nature of the affected bedrock fabric. Earthquakes can submerge large areas of land creating new wetlands. Isostasy accounts for significant changes over hundreds to thousands of years, and allows erosion of a mountain belt to promote further erosion as mass is removed from the chain and the belt uplifts. Plate tectonic movements give rise to orogenic belts and large mountain chains. These mountain chains are in turn responsible for high rates of fluvial erosion and hill slope processes.

1.2 Study and Drawing of Block Diagrams of Important Geomorphological Models

Geomorphological process defines the formation of different landforms on the earth's surface, and denotes cycle of erosion. Landforms evolved with time through series of stages from youth to maturity and to old stages and each of which is characterized by distinctive features. The different Geomorphological models described below:

1.2.1 V shaped valley (Young river valley)

1. The geomorphologic model denotes young stage of a river valley.
2. It is also known as young river valley or V shaped valley, because of its resemblance to English letter V.
3. Young river valley is produced by rapid down cutting and deepening so that stream floor remains comparatively narrow.
4. In youthful stage river flows through hilly regions. Hence, the river suffers heavy headword erosion and develops valleys.
5. Waterfall, Step fall, Gorges are formed in this stage of river based on field condition.

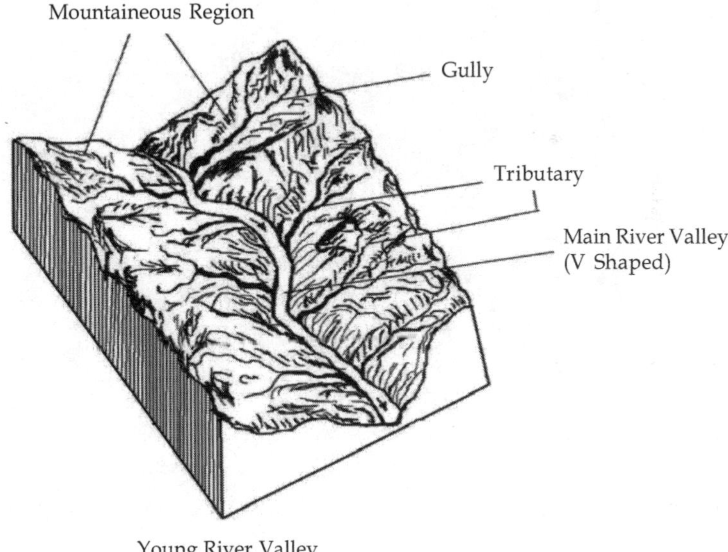

Young River Valley

1.2.2 Mature stage of river

i) The geomorphologic feature of mature river valley is very unique. In this stage river approaches from hilly regions to a flat plain. Hence, there is a sudden drop of flow and the river channels become broader.

ii) A complex branching system is developed. In this stage the river flows in an almost uniform gradient. The drop in gradient reduces the velocity which in turn decreases its erosive power.

iii) With reduced velocity, the river moves in zigzag manner. This is called "meandering of river".

iv) At the inner and upstream side, the velocity of river is low, while at the outer and downstream sides it is greater. Thus deposits are formed on the banks, which in turn form ox-bow lakes.

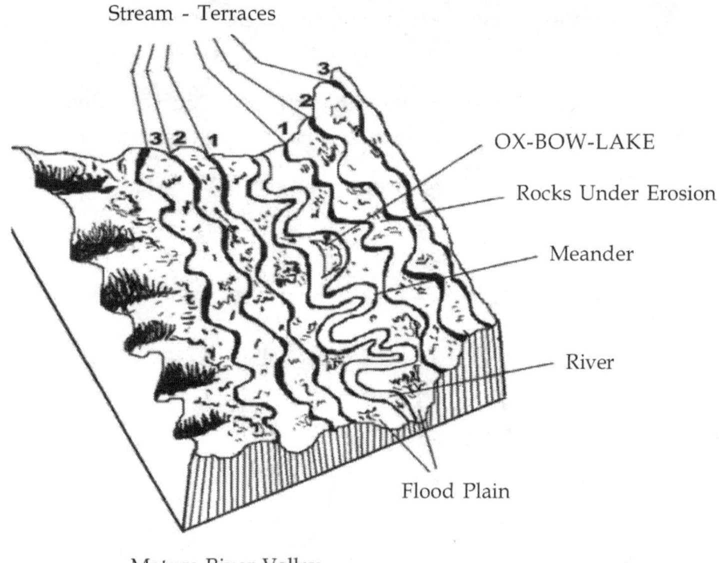

Mature River Valley

1.2.3 Old stage river

i) At the old stage of river, when it reaches to sea, the gradient becomes very gentle and velocity of flow becomes less. Hence, the river loses its erosive power and deposits the transported material and forming many distributaries and ultimately reaches to ocean.

ii) Due to deposition of transported load a broad-depositional landform develops. The structure formed is known as Delta. It is normally funnel or conical in shape from the top.

iii) Delta's can be classified based upon their plan view geometry like:

a) High destructive Delta b) Tidal influenced Delta

c) High constructive Delta d) Lobate (fan) Delta and

e) Bird's foot Delta

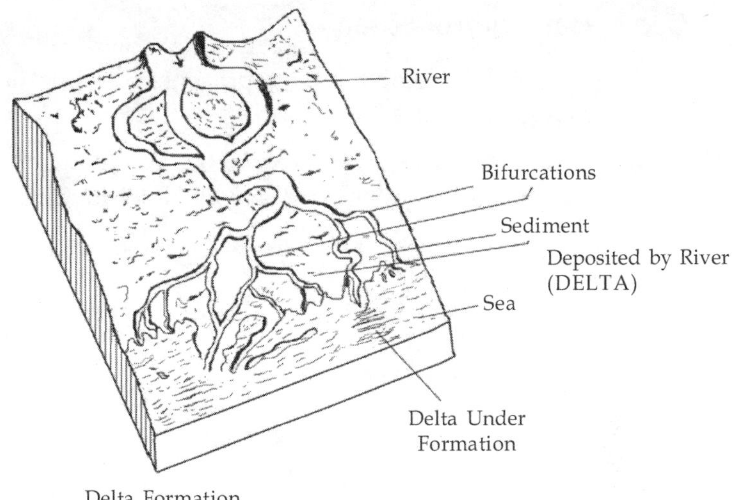

Delta Formation

1.2.4 Caldera

i) The given geomorphologic model represents Caldera lake.

ii) When a volcano becomes dormant, the conical structure collapses and forms a depression that is known as Caldera.

iii) It covers an area of few meters to a kilometer.

iv) Initially, volcanic ash deposits around the volcano pipe and form a volcanic cone. Later when the volcano becomes dormant, it collapses and forms a Caldera.

v) Generally water fills in such depressions and form Caldera Lake.

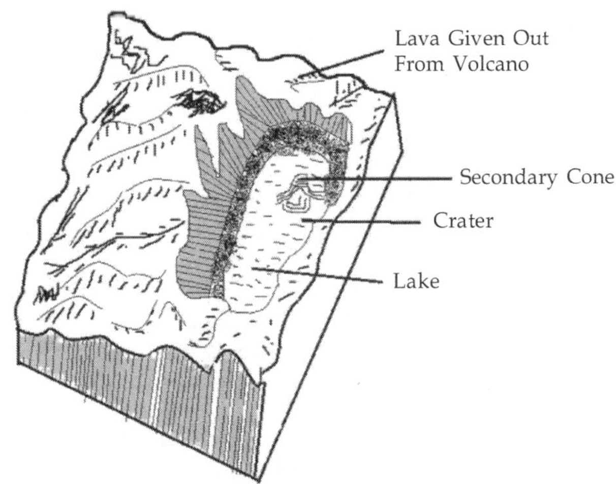

Crater of An Extinct Volcano

1.2.5 Geological action of ocean

i) The given geomorphologic model represents erosional landform, resulted from geological action of ocean.

ii) Due to continuous wave surging action, the less resistant rocks are washed away but the resistant rock forms Notch & Cliff.

iii) At the point of wave attack, the landform is a Notch and the resistant rock form Cliff.

iv) Small outcrop of resistant material inside the ocean form a Island.

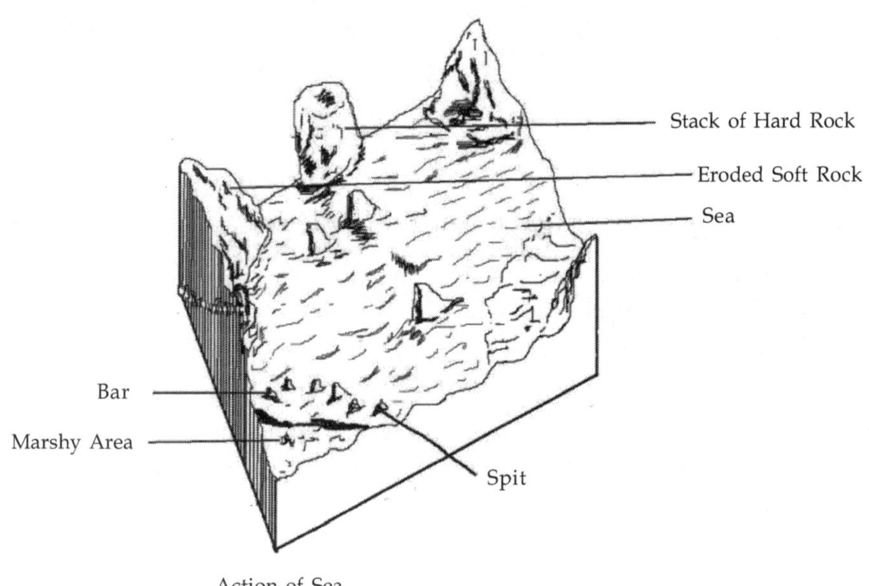

Action of Sea

1.2.6 Groundwater action Karst topography

i) The landscape, which forms in limestone and dolomite rocks as a results of chemical weathering are known as Karst-topography.

ii) In Karstic regions, landform developed by the process of solution activity of groundwater. Limestone is easily dissolved in carbon dioxide rich water (weak carbonic acid) and which is shown in the reaction. $H_2O + CO_2$ ---> H_2CO_3.

iii) Limestone/dolomite is well jointed, allowing water to flow in confined/restricted path. Through the solution action, it dissolves limestone dolomite and reappears at the surface as a stream flow.

iv) The most characteristic feature is sink hole, a conical depression in limestone terrain. When two sink holes side by side merges inside, forms natural bridge.

v) Larger sink holes are called Dolines. They are formed by slow downward movement through solution action, by collapse of roof of an underground cave.

vi) Surface streams in Karstic regions rarely flow longer distance. The point at which a surface steam disappears underground in limestone terrain, is called swallow hole.

vii) In Karstic terrain, sometimes, the underground stream comes to surface and forms vadose, deep phreatic, shallow phreatic spring.

"Karst Topography"

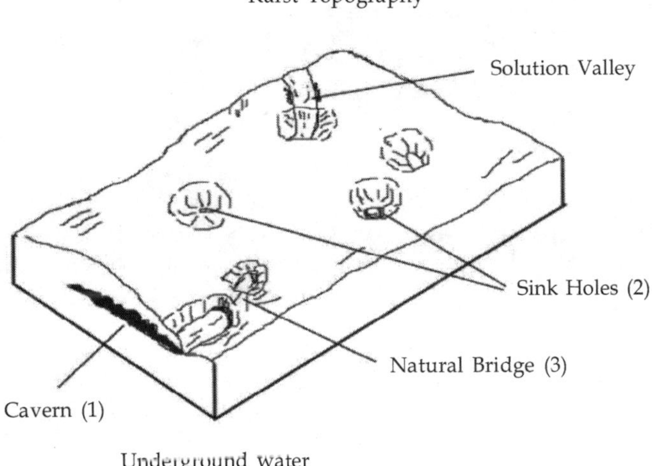

1.2.7 Sand dunes

i) Dunes are small hills of wind-blown sand in desert terrain.

ii) The typical Barchan Dune looks like English letter C. It is a crescent shaped (1 to 50 meter) in height.

iii) Its convex gentler windward side and extending laterally in to two distal horns show the direction of wind as shown in photograph (Fig.)

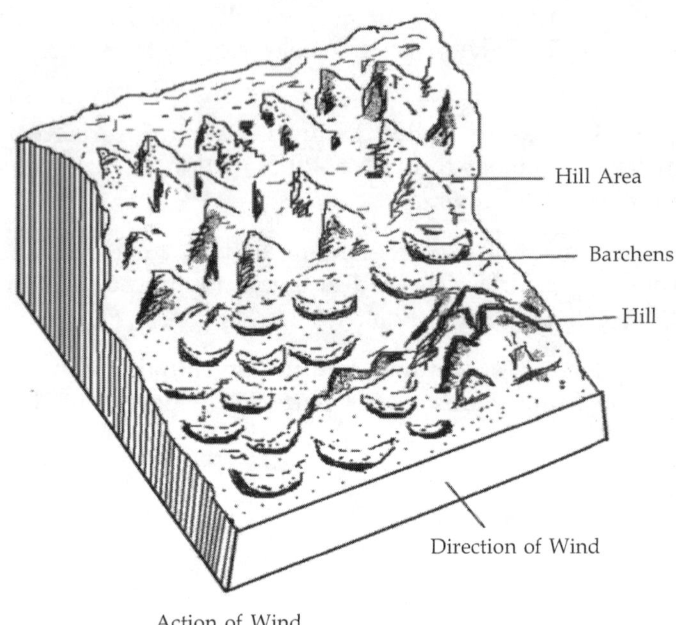

Action of Wind

iv) The other type of dunes are transfers Dune (perpendicular to direction of wind), longitudinal dunes (parallel to wind direction), star dune (rosette) parabolic dunes (U-shape).

❑❑❑

Chapter **2**

Geological Maps

2.1 Introduction

Topographic expression is the replica of geological structures present within the earths surface. When various types of rocks of an area are displayed on a map (sheet of paper), it is known as **Geological Map**. Normally geological maps prepared on topographic maps. Surface topography is best presented by contour lines on a plain paper.

The importance of geological maps is to understand the nature of structure and distribution of surface outcrops with their relation to contour lines. Horizontal and dipping beds are easily identified on map because, horizontal beds runs parallel to contour lines while inclined beds cut through different contours. The angle and direction of dip can be calculated from geological maps. Geological maps are of great importance to civil engineers, after studying geological maps and consultation with geologists, civil engineers can advice for construction of engineering structures like excavation of road cuttings and construction of canal, bridges, dams and tunnels etc. Geologists get vital clues about the geological history of an area, and can draw a cross section of an area and know the structure present underground.

2.2 Terminologies

There are few terminologies used in geological mapping which are given below.

2.2.1 Outcrop

Any rock exposed on the surface of each is known as **outcrop**. It can be depicted on map, when the rock bed elevation equals the surface elevation. For the construction of geological map, a sound knowledge of contour & strike (perpendicular to dip) is essential. Inclination of rock strata is known as Dip.

2.2.2 Contour lines

Contours are used to express surface topography. It is an imaginary line joining points of equal altitude. Mean sea level (MSL) is used as datum

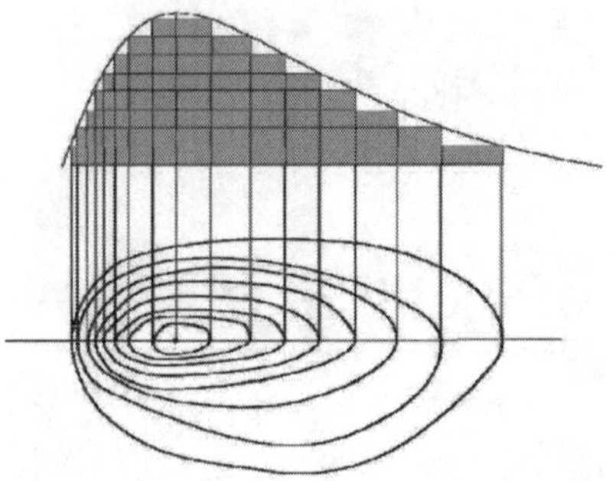

Fig. 2.1 Contour line showing the asymmetrical hill, Horizontal line represents mean sea level.

line/references point for the measurement of heights. Topographic contour lines and strike lines (structural contour) are interpreted similar ways and can be discussed together. As it is clear from the Fig. 2.1 that when contours are at closer intervals they show steep slope, and when wider they show gentle slope. Likewise strike lines represent steep dip, when closer and shows gentle dip. when wider placed. When contours are equally spaced, concentrically arranged, represents isolated rounded hill, while V shaped contours show valley and ridge structure.

2.2.3 Cross section of a contour map and its relation with outcrop and geological structures

Cross section represents the form of topography along a line. For construction of a cross section along point AA' in (Fig 2.2.) First fold the map along A-A' and draw a line on a plane sheet of paper. Points on the map where the contour line intersects A-A' can easily be transferred to the section as shown in (Fig. 2.2). According to scale each contour point is projected to its value, then the cross section is prepared by joining consecutive points, thus obtained (called profile section).

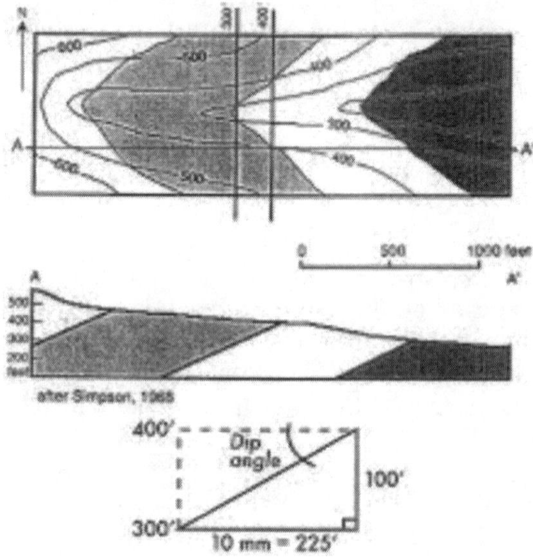

Fig. 2.2 Relation of contour with strike and construction of cross section

2.2.4 Strike line/Structural contour

It is an imaginary line joining points of equal height a dipping bed. When contour line and strike line of a rock formation, it forms of same value are intersecting an outcrop. All strike lines run parallel to each other, and has a value as contours. Dip of a geological formation is perpendicular to strike lines in the direction of their decreasing value. (Fig. 2.3) Line perpendicular to the strike, we pour water on the inclined plane, it would flow in the direction of true dip.

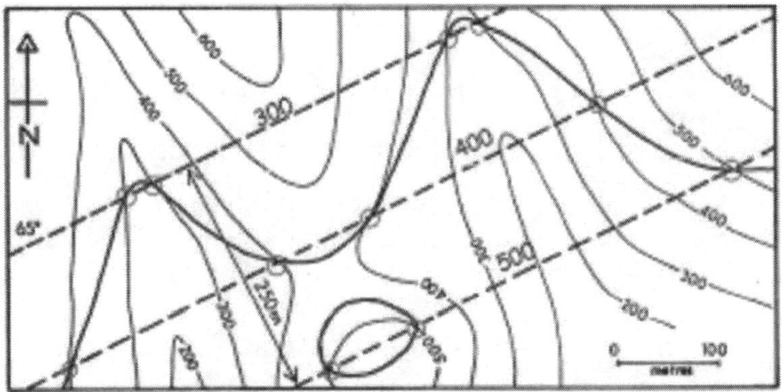

Fig. 2.3 Construction of strike lines of a bed.

2.2.5 Horizontal and dipping bed

The horizontal layered structure formed by deposition of sediments in sheets (beds) and generally have larger areal extent compared to their thickness are known as **Horizontal beds**. In nature, initially the bedding structure is horizontal, but later due to tectonic movements, the beds become inclined and said to be **dipping**. Folds and faults are common deformational structures in tectonically disturbed areas.

Dip : Any rock bed which is disposed at an angle from the horizontal plane, it is called dipping bed. Thus beds are known as dipping when they are not horizontal. Dip measurements have two parameters:

a) **The Direction of dip :** It is the direction in which the bed is inclined. It is measured either in whole circle reading with reference to north, which is taken as 0 to 360 degrees or in quadrant measures which are NE, NW, SE and SW.

b) **Amount of dip :** It is the angle of inclination of bed with horizontal plane.

Dip can be of two types: **True dip and Apparent dip**. Perpendicular to strike line is the true dip and other than this direction are all apparent dips. The value of apparent dip is always less than true dip.

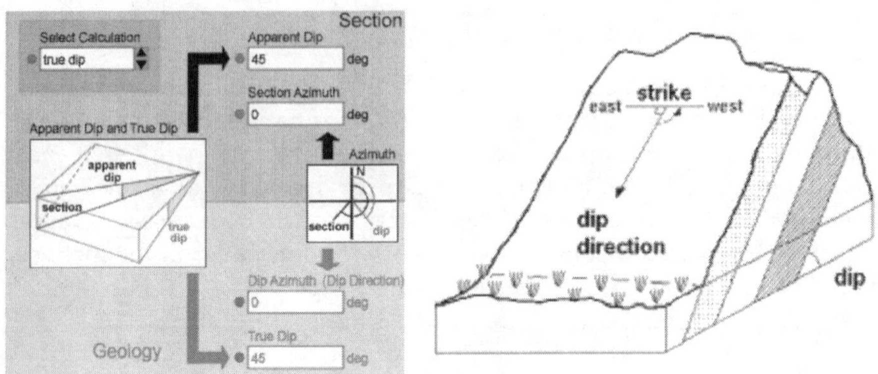

Fig. 2.4 Relation of true and apparent dip

2.2.6 Relation of V shaped outcrop with dip, topographic contours

In geological maps, V shape outcrops is produced when the beds are exposed along a valley. But it differs in shape depends on the direction of upstream and downstream valley.

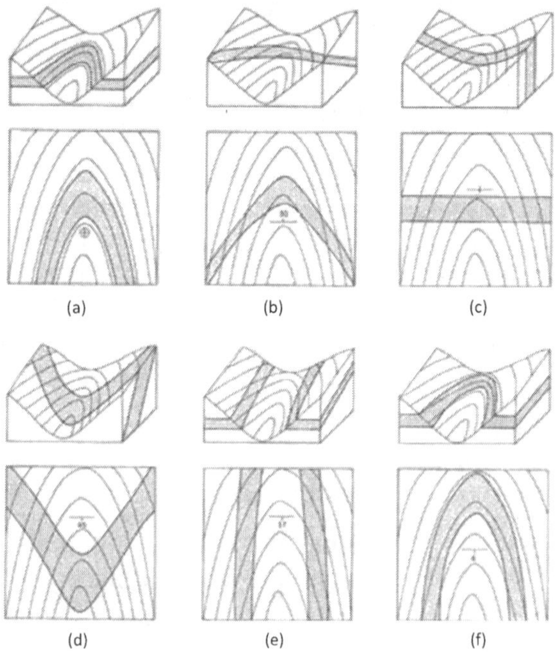

Fig. 2.5 The rule of V's (a) Outcrop patterns of horizontal layer. (b) Outcrop pattern of layer dipping upstream (c) Outcrop pattern of vertical layer (d) Outcrop pattern of layer dipping downstream and angle greater than stream gradient (e) Outcrop pattern of layer dipping downstream at the same angle as stream gradient. (f) Outcrop pattern of inclined layer dipping downstream but with dip less than gradient of the stream.

2.3 Exercises on Structural Geology Problems

There are two types of structural geology problems:

i) Three point problem/ Completion of outcrops and

ii) Drawing and interpretation of cross section from the geological map.

2.3.1 Three point problem/ completion of outcrop

If the contacts of beds (outcrop) are shown at three or more different contour points in a geological map, it is possible to find the direction of strike with the help of three point problem, once the strike line is plotted on map, one can easily complete the outcrop.

For the construction of any line, two points are required. Note the map Number 7 depicts outcrops of a coal seam at A B and C different contour line 200m, 400m & 600m respectively. Join the highest point C (600 m.) to lowest point A (200m). The difference in contour interval is 600-200 = 400m). Assuming the slope is constant we can find a point on AC where the seam is at a height of 400m (mid-point). A straight line drawn through two points represent strike line of 400 metres. We get the trend of strike and now we can construct strike lines of 200 metres through point-A and other with equal strike interval and mark the values of strike line at the border of map. Now we mark a point, representing points same height contour & same height strike and complete the outcrop with by joining all the point with free hand.

2.3.2 Drawing and interpretation of cross section from geological map

In order to illustrate a 3D structure of an area, a map in general is not enough. Therefore, cross-sections of the area may accompany the map. At first, we draw a baseline, exact length of the line X-Y, on map. Then mark the points at which the contour line crosses the line of section and also put the values of contour at the bottom as shown. Depending on scale, mark points vertically above the line, and lastly join all the points with free hand and. Now the topography is prepared as shown in the cross section above XY (Fig. 2.6).

2.3.3 Map exercises with few solutions

Constructing of strike line (Structural contours)

For the construction of strike line, two points are necessary. In the map No. 1 these two points should be marked from any one bed contact

T/S, S/R or R/Q. For example, the boundary between beds S and T on map cuts the 700 m contour at three points. These points lie on the 700 m strike line, join these three points and form the strike line having value 700 m. Now construct another strike lines 600 value in the same contact and mark the value, the dip direction can easily be determined where the strike line show decreasing value. True dip direction is perpendicular to the strike line, while other directions will be said apparent dip. Like topographic construction, mark the points of different contacts on cross section. Now with the help of following formula dip amount is determined.

$$\text{Dip} = \frac{\text{Vertical Interval (VI)}}{\text{Horizontal Equivalent (HE)}} \text{ or } \frac{\text{Contour Interval}}{\text{Strike Interval}}$$

Contour interval is 100 m. and strike interval along section line can be measured according to scale given in map. The ratio between V I and HE would be the dip amount and all the beds would be projected on the top of cross section profile and angles are made as shown in the Fig. 2.6. Map No. 2.7 & 2.8 shows the relation of topography with contours. Map No. 4 to 8 are exercises on three point problem Map No. 9 is drawing and interpretation of cross section map with solution. Map No. 10 and 11 are problems on cross section Map.

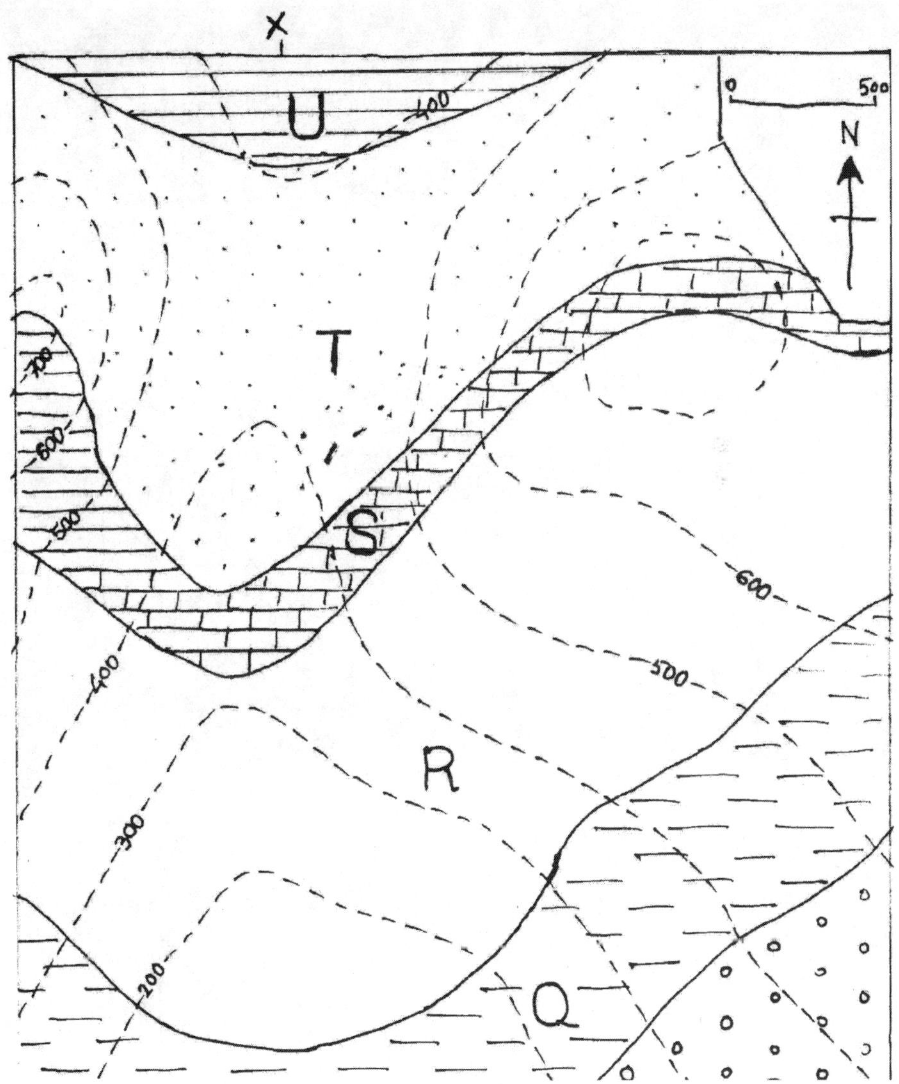

Map 1. Cross Section Map

Problem : Draw a cross section along X-Y and describe the geology of the area.

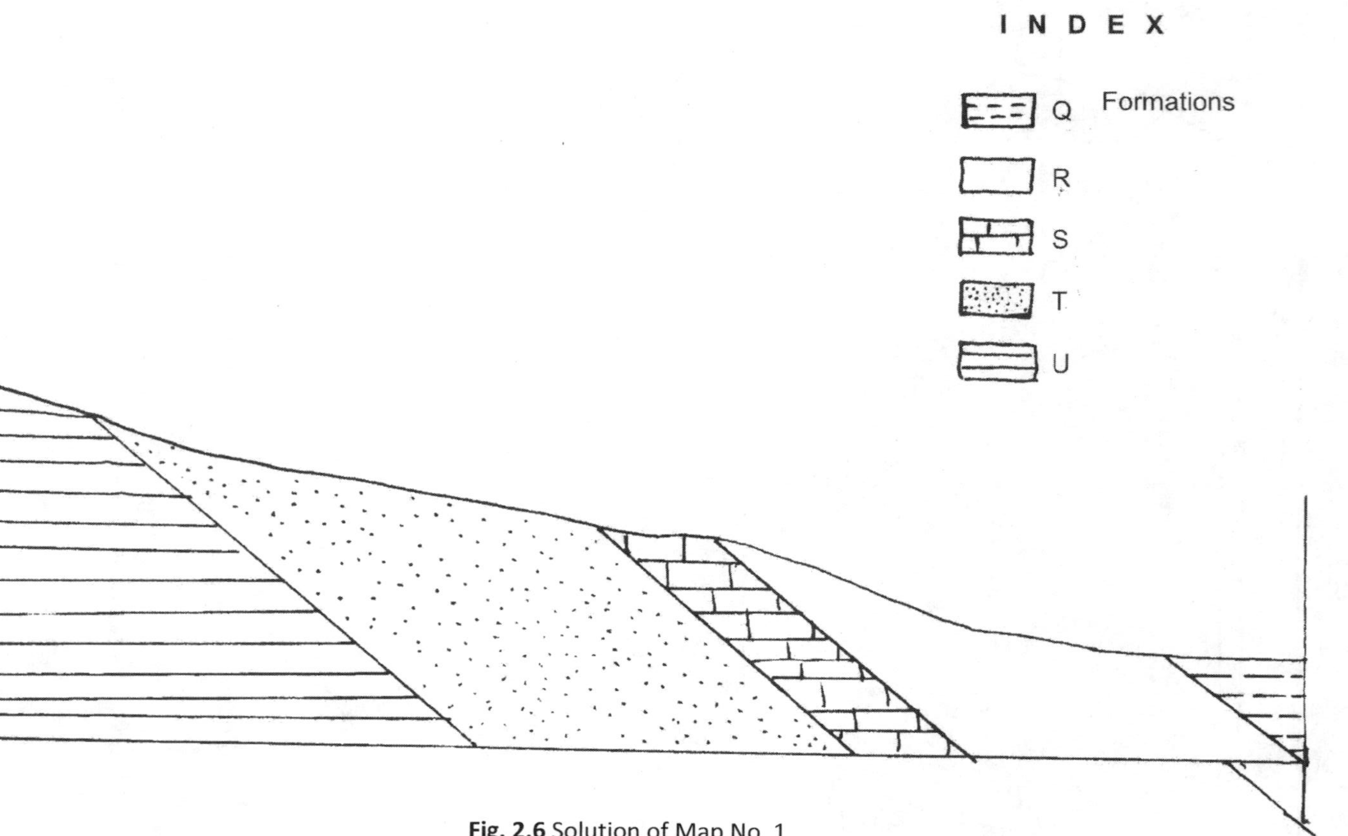

Fig. 2.6 Solution of Map No. 1

Interpretation of Contur Maps

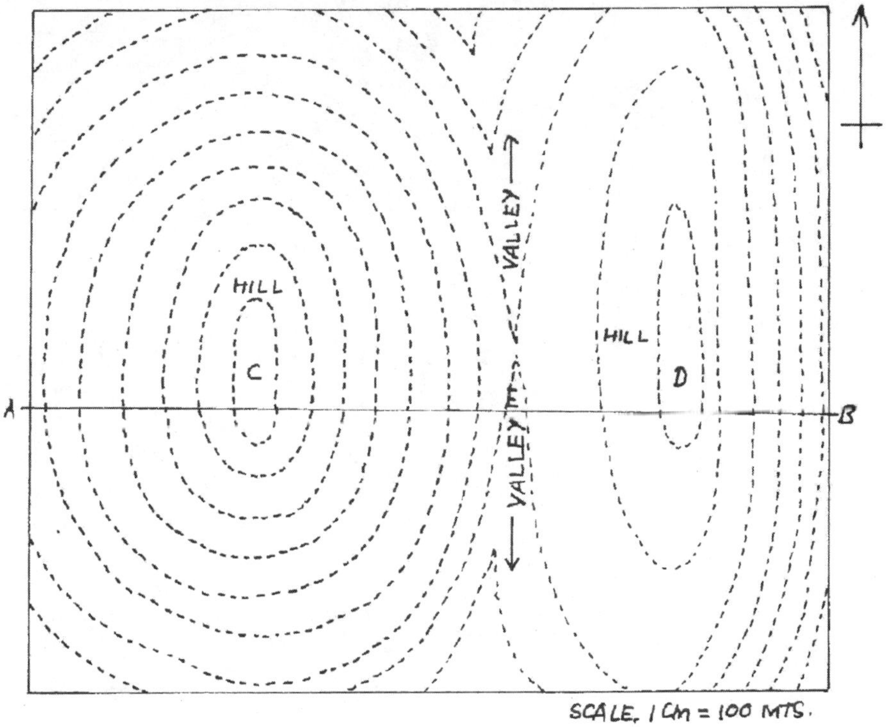

SCALE. 1 Cm = 100 MTS.

Topography along line AB

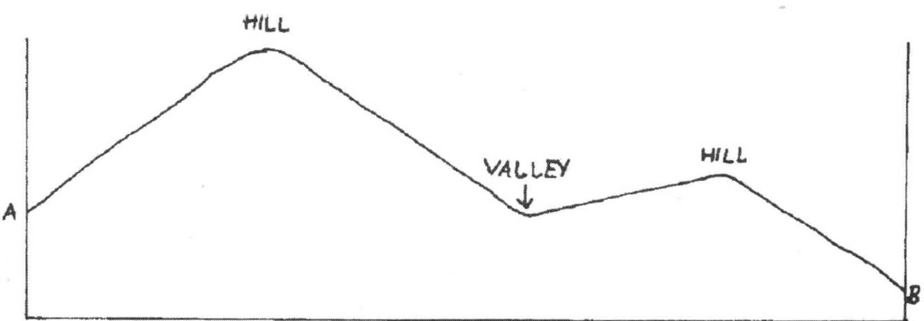

Map 2.7. Interpretation on contour map and Topographic map

Interpretation of Contur Maps

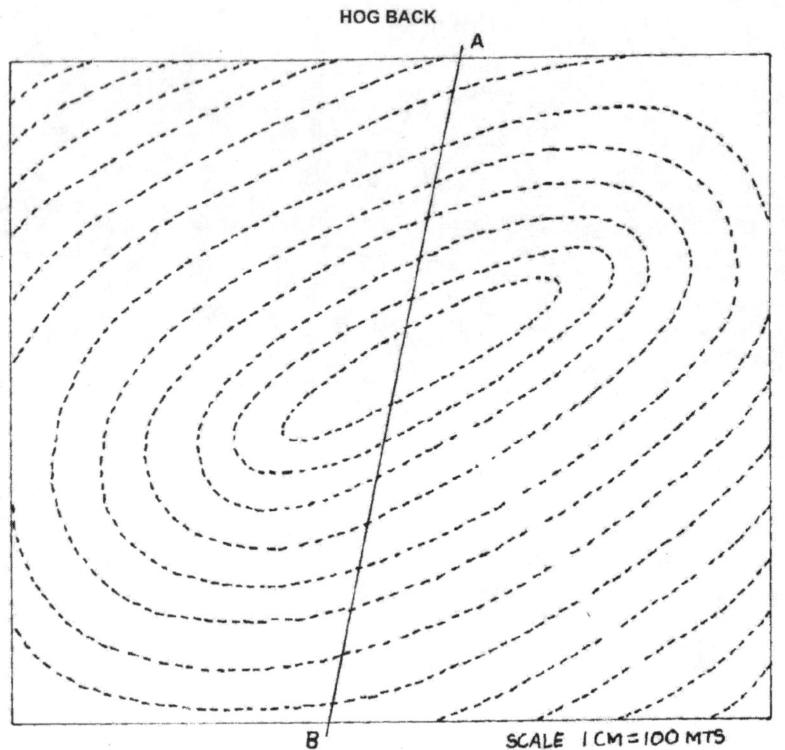

Topography along line AB.

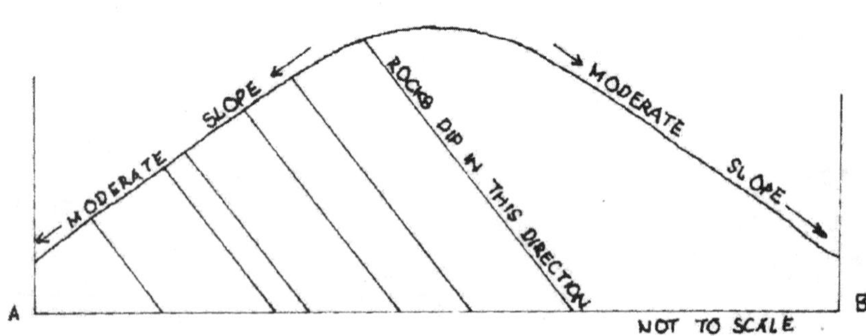

Map 3. Interpretation of contour map

Three point proble - Based on three outcrops in different height points complete the outcrop in both the maps

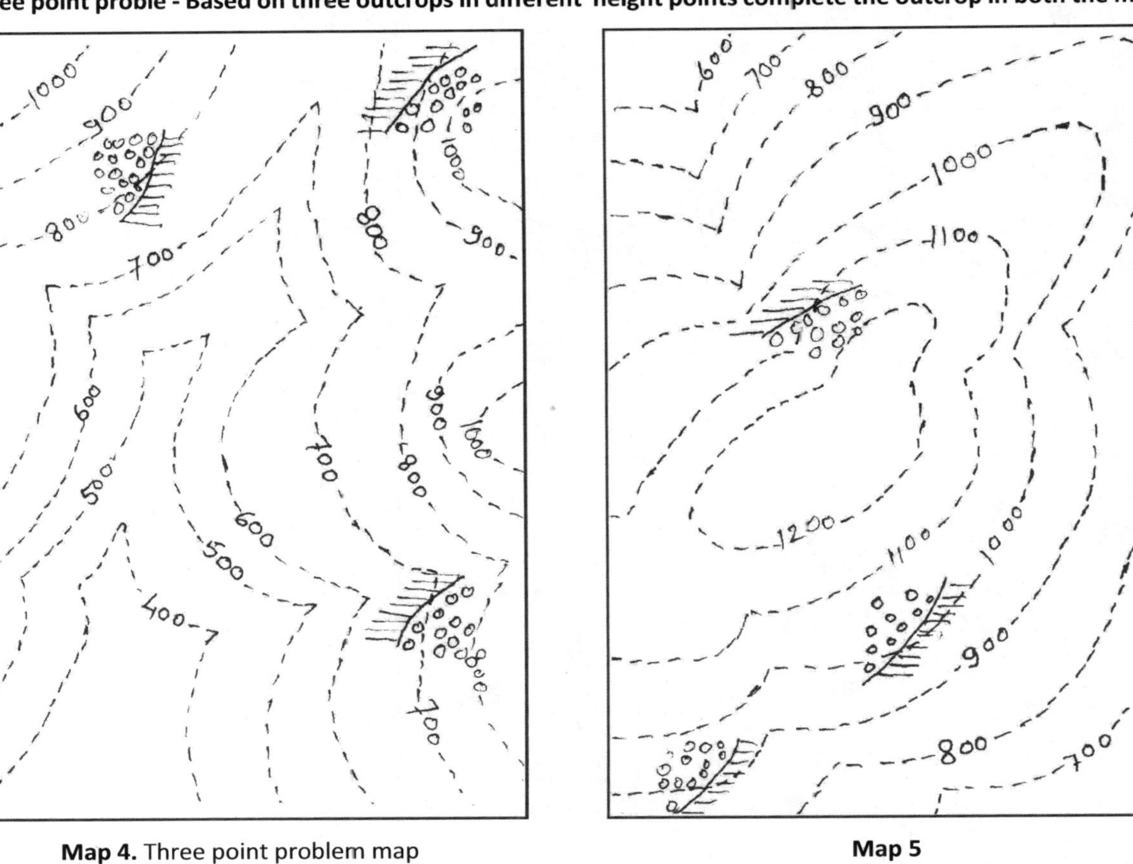

Map 4. Three point problem map

Map 5

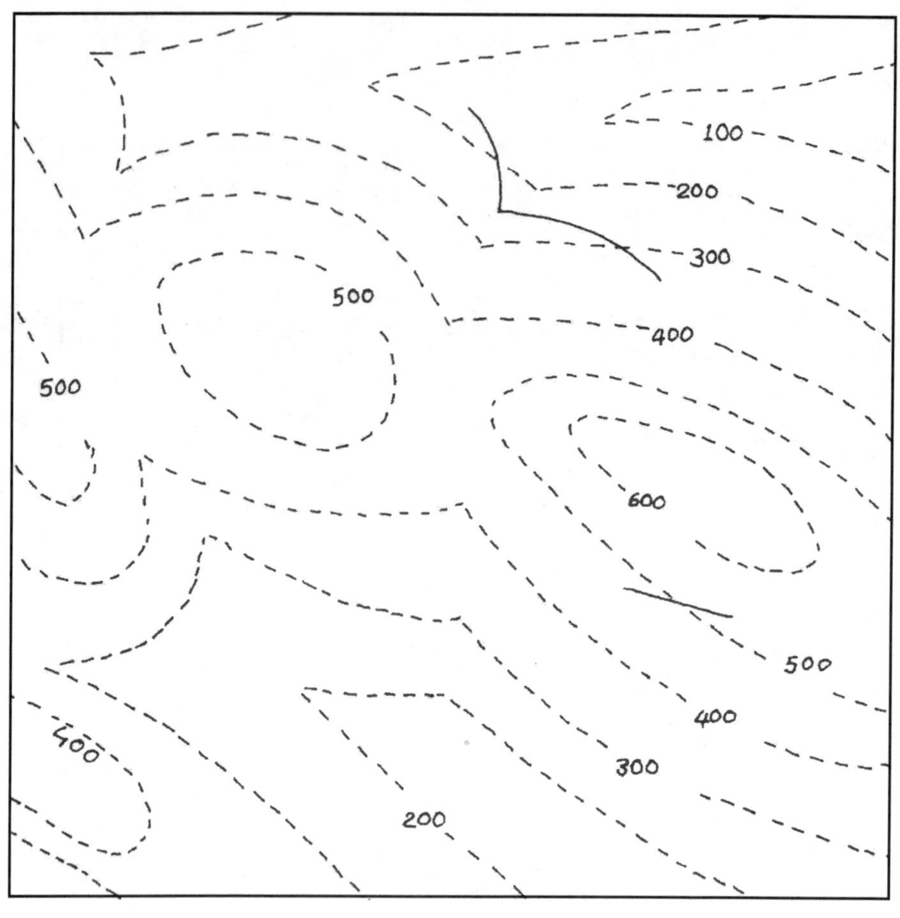

Completion of Outcrop

Problem : Draw strike lines based upon the outcrops shown at contour 200, 300 and 500 ft. and complete the outcrop.

Map 6

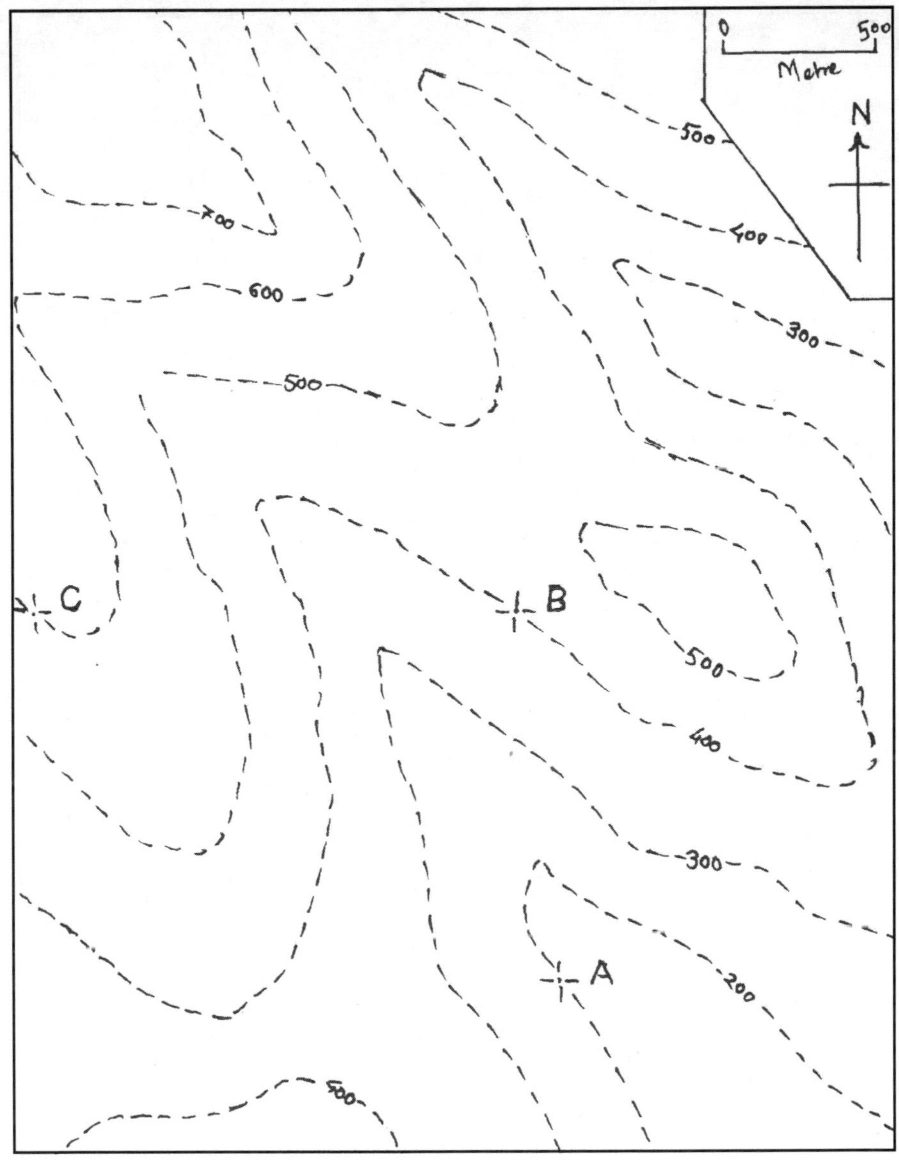

Completion of Outcrop

Problem : Based on three outcrops at point AB & C complete the map.

Map 7

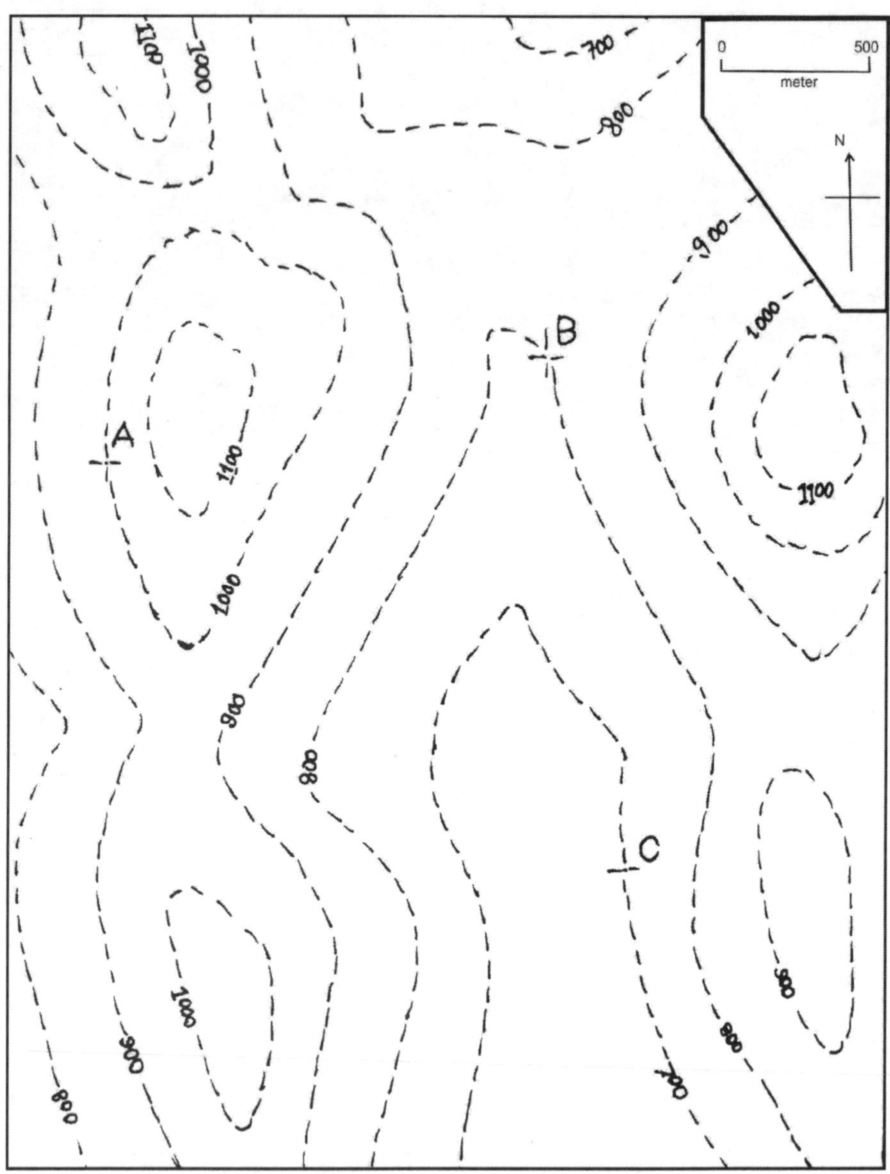

Completion of Outcrop (Three Point Problem)

Problem : Draw strike line based on the information given in the map. A,B, C are the outcrops. Complete the map.

Map 8

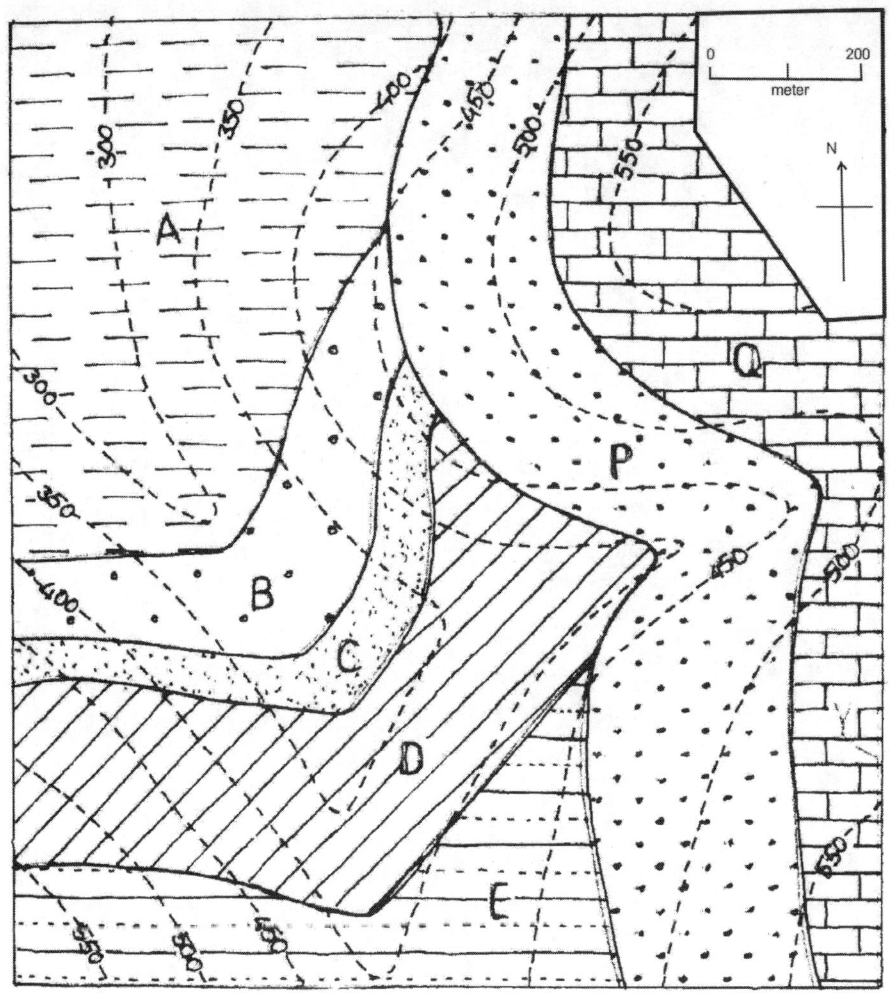

Cross Section Map

Problem : Draw a cross section along X-Y and describe the geology of the area.

Map 9

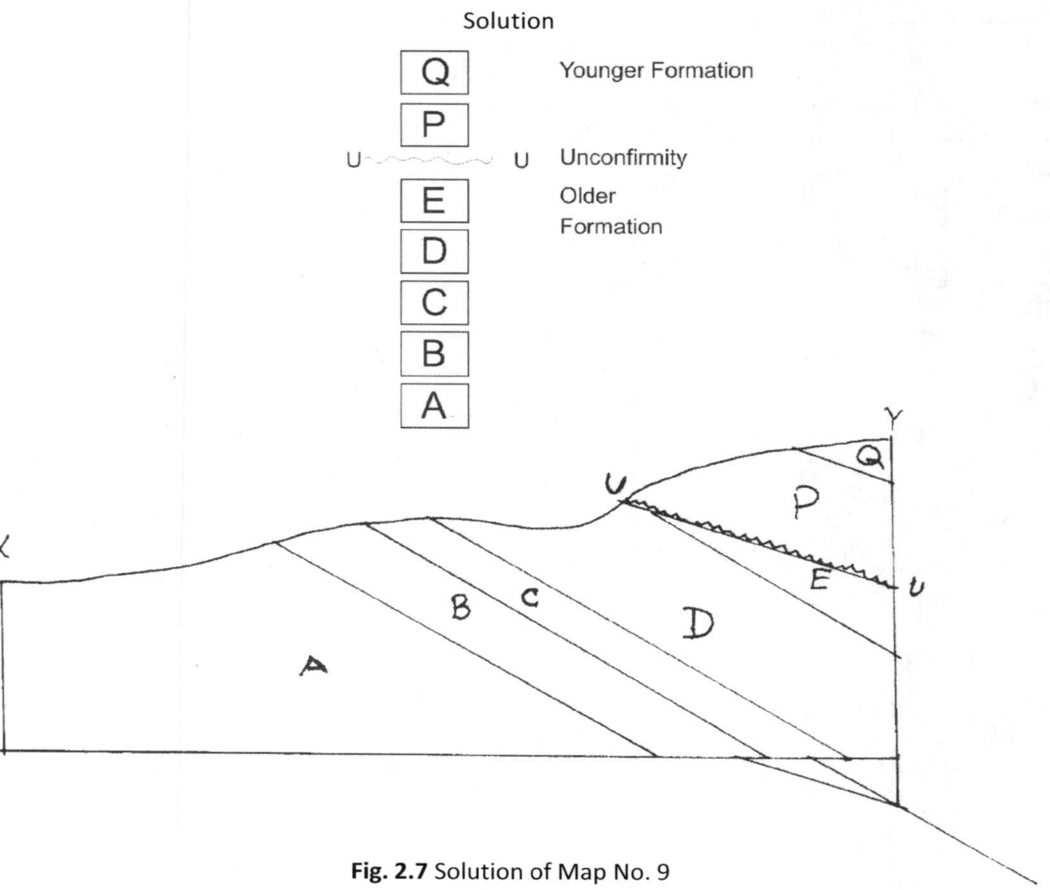

Fig. 2.7 Solution of Map No. 9

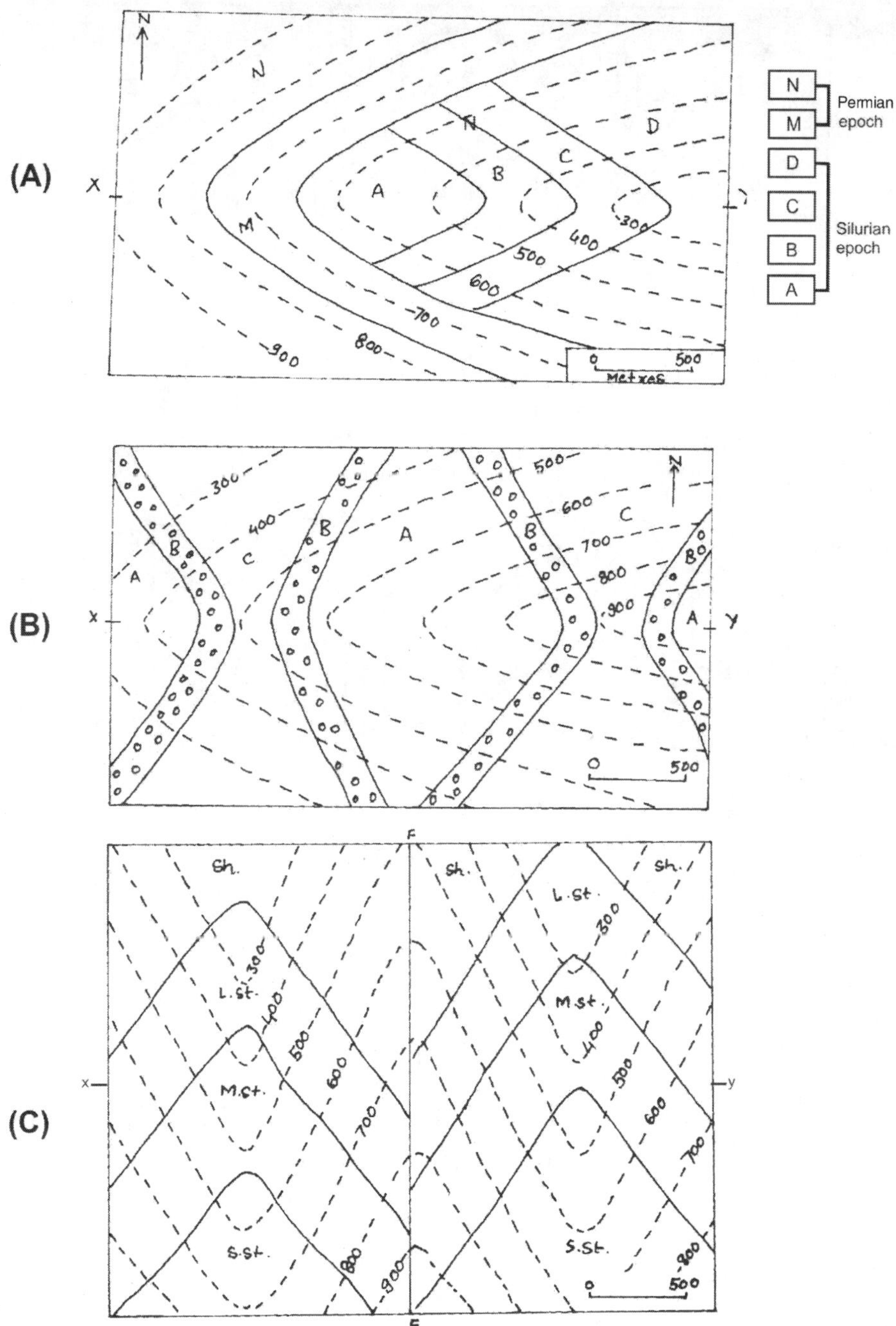

Problem : Draw a cross section along line X-Y and describe the geology of the area map (A), (B) & (C)

Map 10

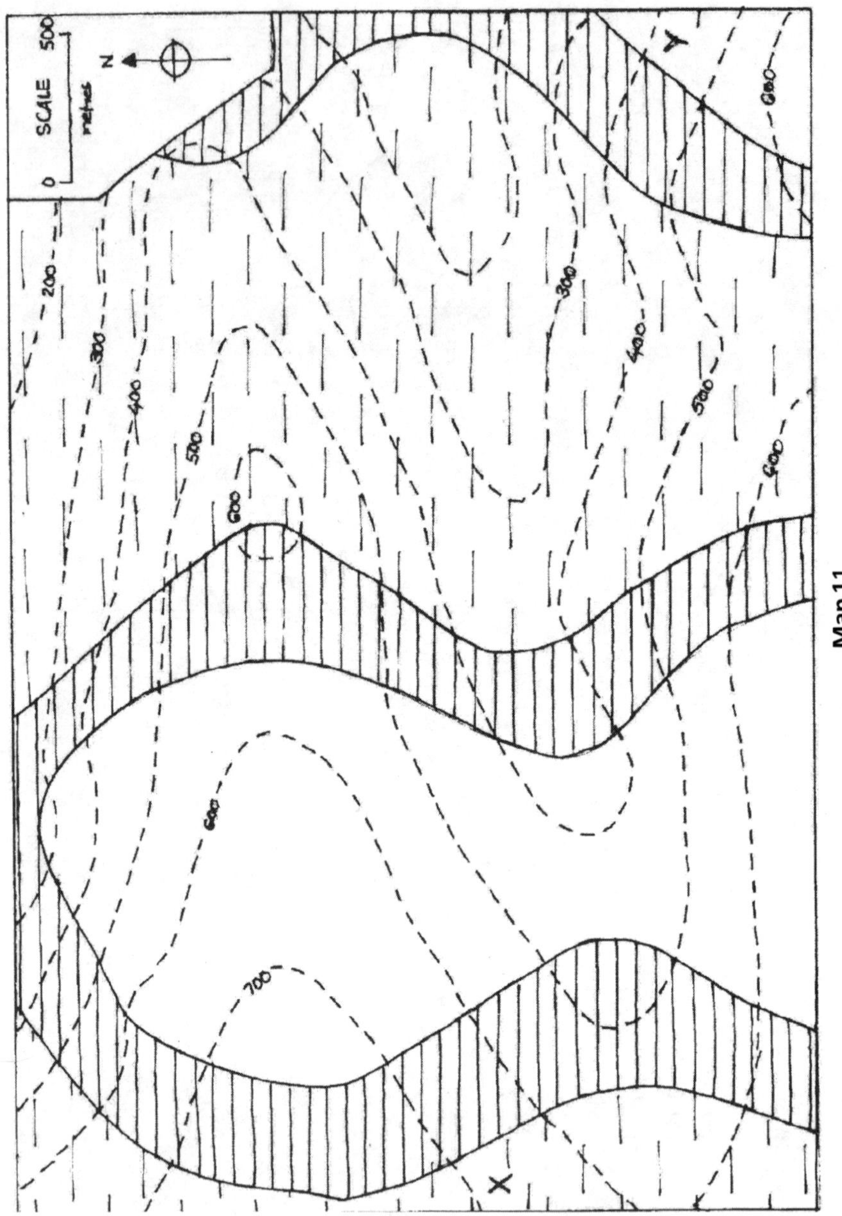

Map 11

Problem: Draw a cross section along the line X-Y and describe the Geology

❑❑❑

Chapter 3

Crystallography

3.1 Introduction

Crystallography is the science of crystals. The term crystal is derived from the Greek word "Crystallos" which means "ice". The term was used for colorless quartz which was believed to be "fossilized" ice. Crystals are bodies bounded by flat surface, arranged on a definite plan which is an expression of the internal arrangement of the atoms. They are formed by consolidation of magma which is also considered as crystallization.

3.2 Terms in Use to Describe a "Crystal"

3.2.1 Faces

Faces are the plane surface in a crystal. The common faces are usually parallel to net planes containing the greatest number of lattice points or ions. Wide spacing between the net planes leads to development of faces in these planes. Faces are two types like and unlike faces.

3.2.2 Forms

A crystal contains entirely of like faces are said to be of simple form. For example cube and octahedron are crystals with **simple forms**. A crystal consists of two or more simple forms is called combination.

3.2.3 Edge

The intersection of two adjacent faces of a crystal is called an edge. It is the reflection of proper internal arrangement of atoms.

3.2.4 Solid Angle

The intersection of three or more faces in a crystal forms solid angle.

3.2.5 Euler's Theorem

The relationship between solid angle, faces and edges of a crystal expressed in a simple formula, given by a scientist Euler known as **Euler's theorem**.

S + F = E + 2 Where: S= Solid Angle, F = Faces, E= Edges

3.2.6 Interfacial angle

The actual measured angle between the normals of two faces of a crystal is known as **interfacial angle**. It is characteristic of a crystal and has great significance in crystallography. In Fig. 3.1 the interfacial angle between two faces is shown by **A**. The interfacial angle is measured by Goniometer. This instrument is of two types : i) Contact Goniometer (Fig. 3.2) and ii) Reflecting **Goniometer**. Contact Goniomenter is used to measure interfacial angles of normal and bigger crystals, but reflection goniometer are suitable for smaller crystals.

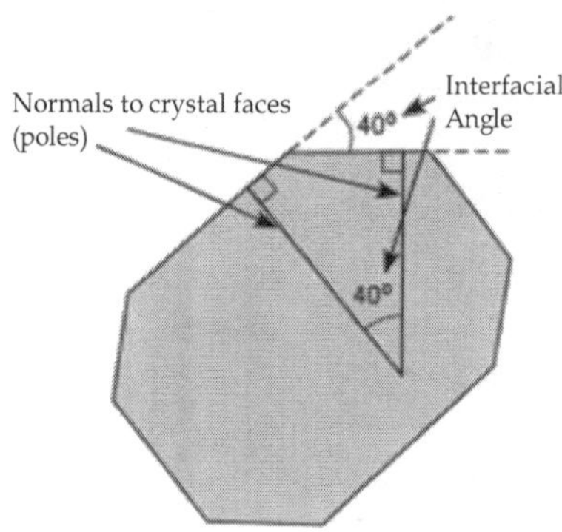

Fig. 3.1 Interfacial Angle

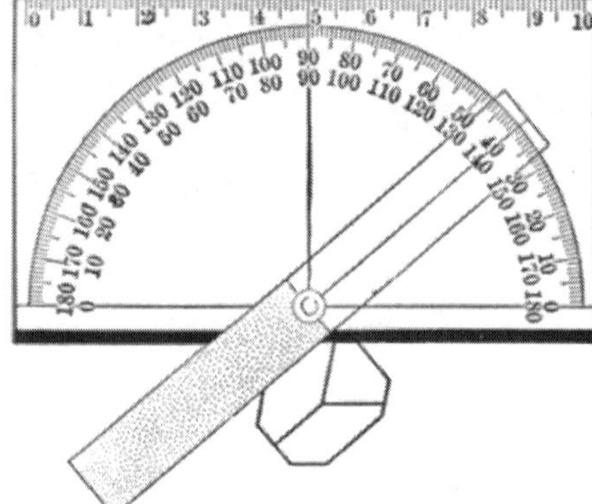

Fig. 3.2 Contact Gonometer

3.2.7 Law of constancy of Interfacial angle

The interfacial angles are always constant for all crystals of a given mineral. They should have same chemical composition and crystallize at the same temperature. This is known as **law of constancy of Interfacial angle**.

3.3 Symmetry

Crystals have regular arrangements of faces edges and solid angles. This regularity is said symmetry of a crystal. The symmetry of a crystal can be best studied with the help of three parameters:

Plane of symmetry: It means a crystal can be divided into different planes that are mirror image of each other. Example: A cube has 9 planes that are mirror images to each other. Fig.3.3

- **Axis of symmetry:** It means how many times faces come when a crystal is rotated 360 degree along an axis. Example Axes of symmetry in a cube are 13. Fig. 3.3

  3 axes show 4 fold symmetry represented as (3A4) or 3iv

  4 axes show 3 fold symmetry represented as (4A3) or 4iii

  6 axes show 2 fold symmetry represented as (6A2) or 6ii

  Total 13 Axes (3+4+6) (Fig. 3.4)

- **Centre of Symmetry:** Centre of Symmetry means when like faces edges arranged in pairs and in opposite sides of centre point. The crystal is said to be having a centre of symmetry.

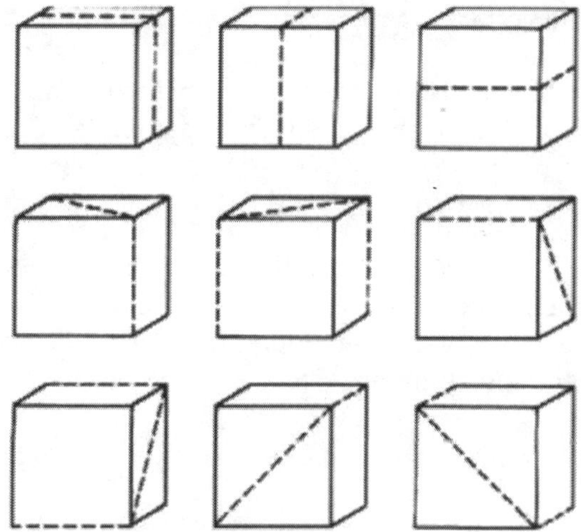

Fig. 3.3 Plane of symmetry

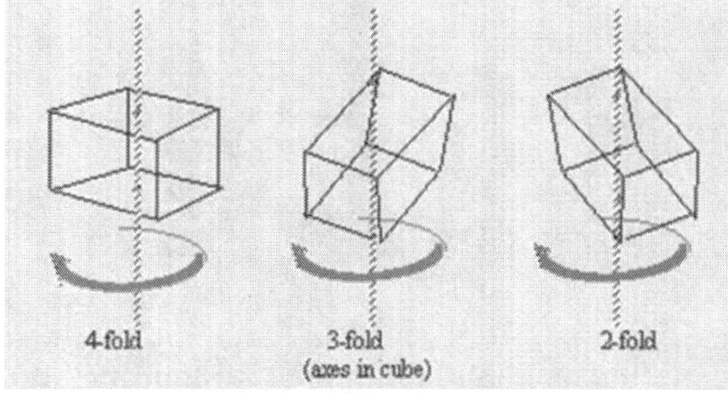

Fig. 3.4 Axis of Symmetry

3.4 Crystal Systems

There are 6 major crystal systems in which most of the minerals crystallize in nature.

3.4.1 Cubic system or Isometric system

The first and simplest crystal system is the isometric or cubic system. It has three axes, all of which are the same length denoted by a_1, a_2 and a3 and all intersect at 90 degrees to each other. Because of the equality of axes, minerals in the cubic system are singly refractive. According to different symmetry elements it can be divided into three types (named after minerals):

- Galena type
- Pyrite type
- Tetrahedrite type

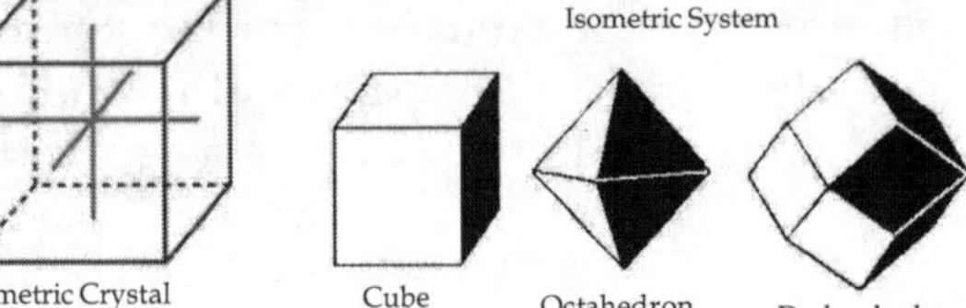

Minerals that form in the isometric system include all garnet, diamond, fluorite, gold, lapis lazuli, pyrite, silver, sodalite, sphalerite, and spinel.

3.4.2 Tetragonal system

The tetragonal system also has three axes which meet at 90 degrees. It differs from the isometric system in that the C axis is longer than the a_1 and a_2 axes, which are of the same length. Mineral Zircon crystallizes in this system, hence the symmetry class is:

Zircon type

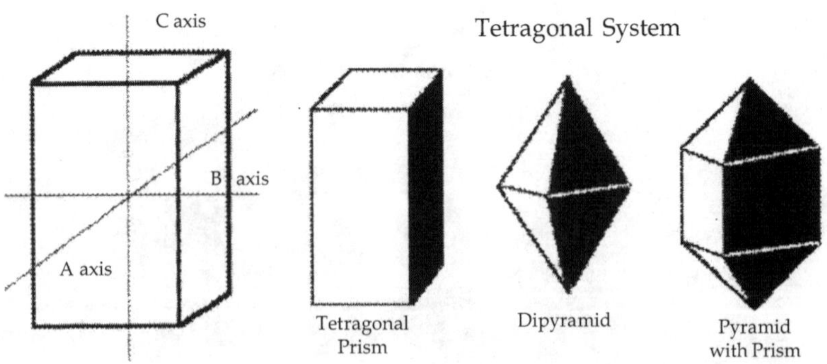

Minerals that form in the tetragonal system include apophyllite, idocrase, rutile, scapolite, wulfenite, and zircon.

3.4.3 Hexagonal system

The crystal systems mentioned above represent every variation of four-sided figures with three axes. In the hexagonal system we have an additional axis, which gives the crystals six sides. Three of these are equal in length and meet at 60 degrees to each other. The C or vertical axis is at 90 degrees to all the shorter axes.

Based on symmetry elements, Hexagonal system is divided into four types:

- Baryl type
- Calcite type
- Tourmaline type
- Quartz type

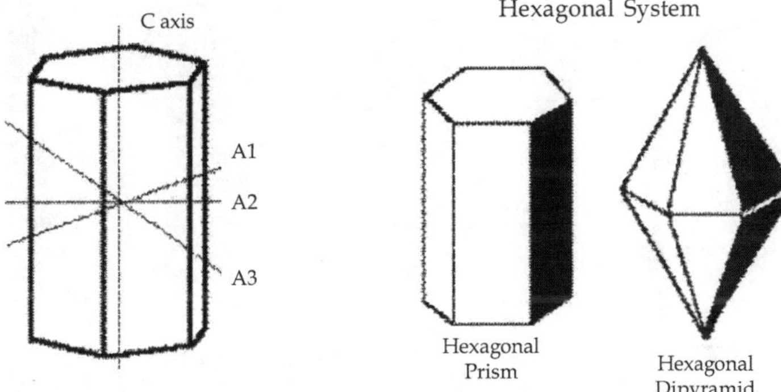

3.4.4 Orthorhombic system

The orthorhombic system has three axes, each of which is a different length. These axes intersect at 90 degree angles. Orthorhombic System have three axis arranged, 90 degrees to each other, but are different in lengths. Barite crystallizes in this system, hence the symmetry class is called Baryte type.

Baryte type

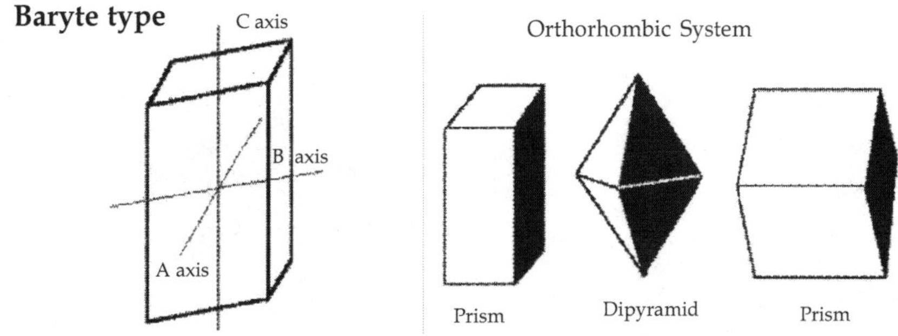

Minerals that form in the orthorhombic system include andalusite, celestite, chrysoberyl (including alexandrite), cordierite, epidote, enstatite, hemimorphite, fibrolite/sillimanite, hypersthene, olivine, sulphur, topaz, and zoisite.

3.4.5 Monoclinic system

In the monoclinic system, two of the axes, A and C, meet at 90 degrees, but Axis B does not. All axes in the monoclinic system are different lengths. Gypsum crystallizes in this system so the symmetry class is known as Gypsum Type

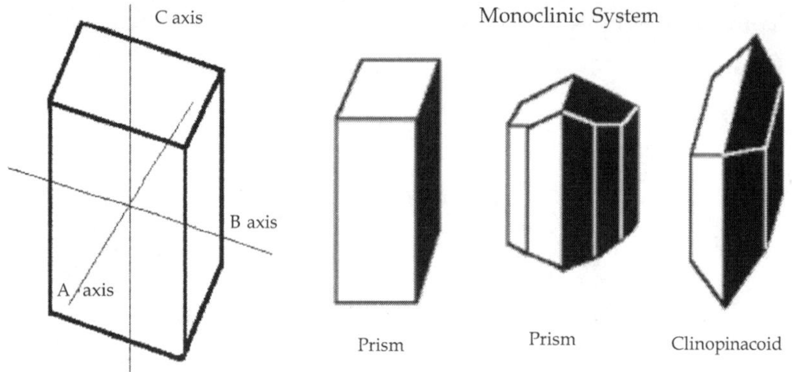

Minerals that form in the monoclinic system include azurite, diopside, jadeite, lazulite, malachite, orthoclase feldspars (including albite moonstone), staurolite, sphene, and spodumene.

3.4.6 Triclinic system

In the triclinic system, all the axes are different lengths and none of them meet at 90 degrees. Feldspar forms are good example of triclinic crystal system. The symmetry class is known as Axinite Type.

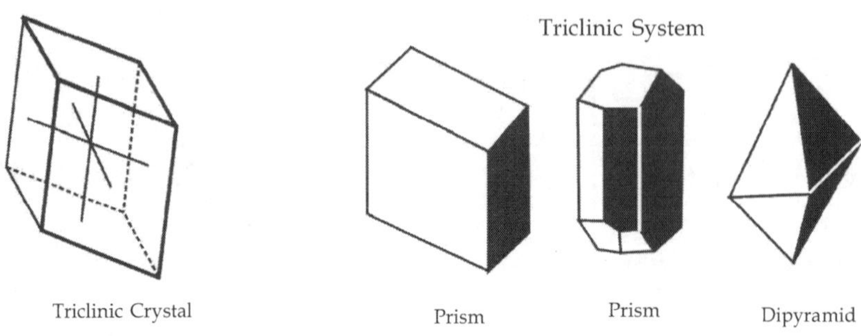

Minerals that form in the triclinic system include amblygonite, axinite, kyanite, labradorite, microline feldspar, plagioclase feldspars (including labradorite), rhodenite etc.

3.5 Description of Some Common Minerals Crystallizes in Different Crystal Systems

3.5.1 Zircon, Rutile and Visuvianite

a) The given crystal belongs to Tetragonal System, as it contains two equal horizontal axis and third vertical long axis.

b) The crystal contains following faces :

- Tetragonal Prism of Ist Order

  4 faces Miller Symbol 110

- Tetragonal Pyramid of Ist order

  8 faces Miller Symbol 111

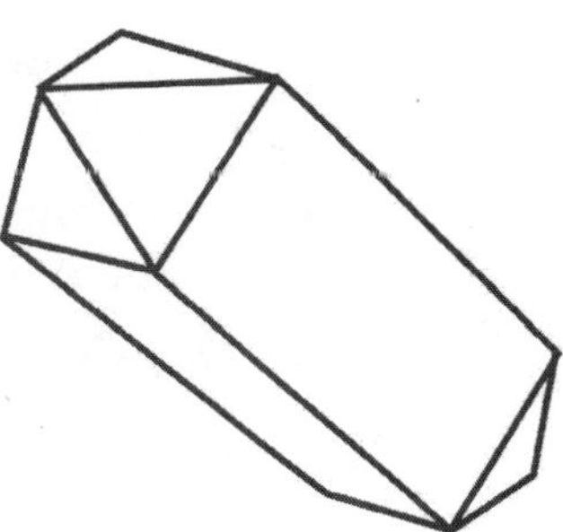

c) Therefore the given crystal is of Zircon.

3.5.2 Baryte

The given crystal belongs to Orthorhombic system as all the axis are unequal in length.

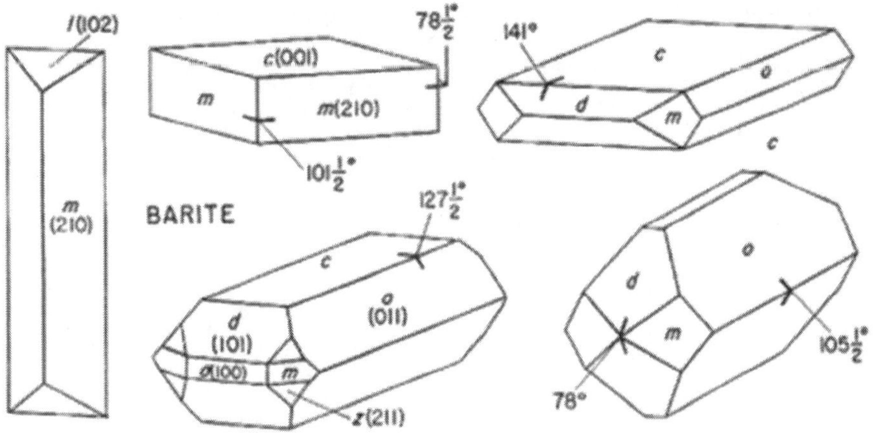

PINACOIDS *a*, *c*, PRISMS *m*, *o*, *d*, *l*, DIPYRAMID *z*.

a) The crystal contains following faces:
   - Basal Pinacoid (1) 2 faces Miller Symbol 001
   - Prism (2) 4 faces Miller Symbol 110
     - Brachy dome (3) 4 faces Miller Symbol 101
     - Macrodome (4) 4 faces Miller Symbol 011

b) The given crystal is of **Baryte**.

3.5.3 Olivine

a) The given crystal model belongs to orthorhombic system. It contains unequal axis a b and c.

b) Crystal model contains following faces: (Identification Number mentioned on the figure)
   - Basal pinacoid (1) 2 faces Miller Symbol 001
   - Macro pinacoid (2) 2 faces Miller Symbol 100
   - Brachy pinacoid (3) 2 faces Miller Symbol 010
   - Prizm (4) 4 faces Miller Symbol 110
   - Macrodome (5) 4 faces Miller Symbol 101
   - Brachy dome (6) 4 faces Miller Symbol 011
   - Pyramid (7) 8 faces Miller Symbol 111

c) Based upon above description the given crystal model is of **Mineral Olivine**.

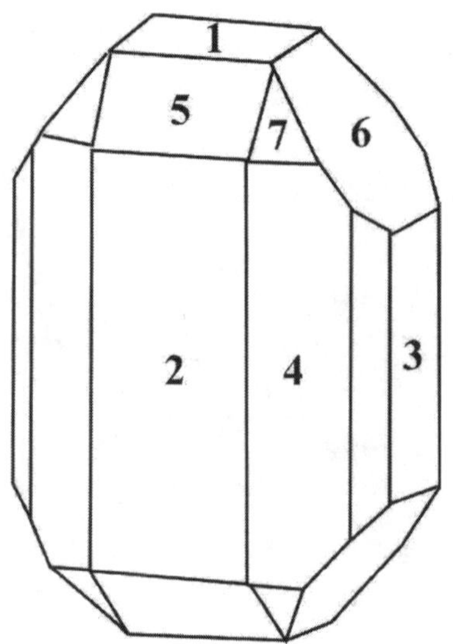

3.5.4 Staurolite

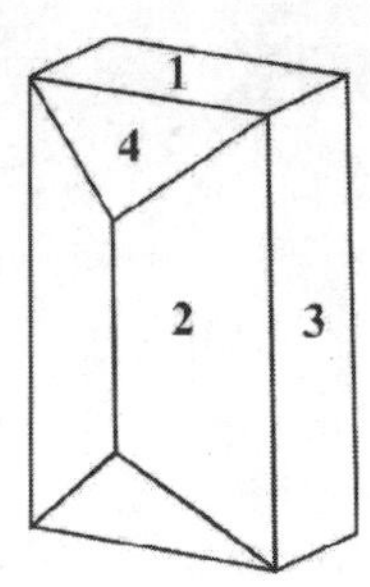

a) The given crystal model belongs to orthorhombic system. It contains unequal axis a b and c.

b) Crystal model contains following faces: (Identification Number mentioned on the figure)

- Basal pinacoid (1) 2 faces Miller Symbol 001
- Macro dome (4) 4 faces Miller Symbol 101
- Brachy pinacoid (3) 2 faces Miller Symbol 010
- Prizm (2) 4 faces Miller Symbol 110

c) Based upon above description the given crystal model is of M**ineral Staurolite**.

3.5.5 Gypsum

a) The given crystal model belongs to monoclinic system. It contains unequal axis a b and c. in which a axis is inclined.

b) Crystal model contains following faces: (Identification Number mentioned on the figure)

- Clino pinacoid (1) 2 faces Miller Symbol 010
- Negative Hemipyramid (3) 4 faces Miller Symbol 111
- Prizm (2) 4 faces Miller Symbol 110

c) Based upon above description the given crystal model is of **Mineral Gypsum**.

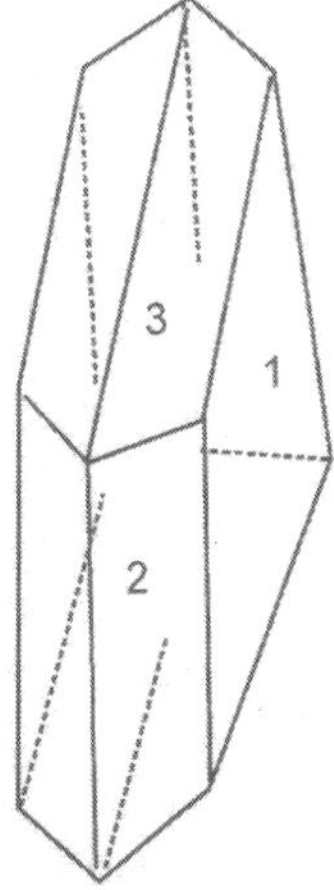

3.5.6 Orthoclase

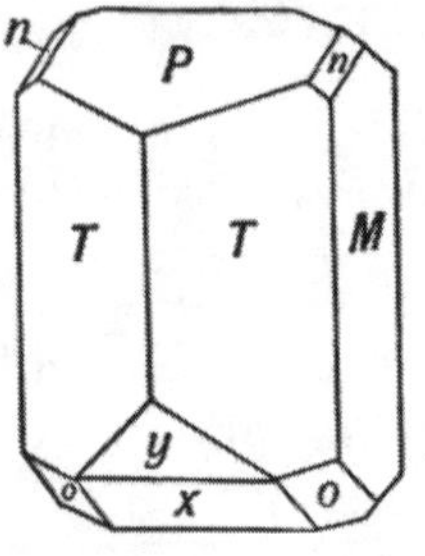

a) The given crystal model belongs to monoclinic system. It contains unequal axis a b and c in which a axis is inclined.

b) Crystal model contains following faces: (Identification Number mentioned on the figure)

- Basal pinacoid (1) 2 faces Miller Symbol 001
- Clino pinacoid (2) 2 faces Miller Symbol 010
- Prizm (3) 4 faces Miller Symbol 110
- Prizm (4) 4 faces Miller Symbol 130
- Positive Hemipyramid (5) 4 faces Miller Symbol 111
- Positive Hemipyramid (6) 4 faces Miller Symbol 201
- Clinodome (7) 4 faces Miller Symbol 011

c) Based upon above description the given crystal model is of **Mineral Orthoclase**.

❏❏❏

Chapter **4**

Minerals and Rocks

The earth's crust is made up of minerals and rocks. Minerals are inorganic substances with a definite chemical composition, whereas rocks are mixture or aggregates of minerals. Each mineral is composed of certain elements present in fixed proportion. For example, mineral quartz, SiO_2 is a compound of the elements silicon (Si) and oxygen (O) combined in the proportion of 1:2, so its chemical formula is SiO_2, whereas rock granite is a mixture of several minerals like quartz, feldspar etc. Nature has provided unique physical properties to different minerals that can be used to identify them.

4.1 Physical Properties of Minerals

Majority of minerals contain following properties that can be used to identify them in field and in laboratory: crystal habit, cleavage, hardness, density, luster, streak, color, tenacity, magnetism, and taste.

4.1.1 Crystal habit

The faces that develop on a crystal depend on the space available for the crystals to grow. If crystals grow into one another or in a restricted environment, it is possible that no well-formed crystal faces will be developed. However, crystals sometimes develop certain forms more commonly than others, although the symmetry may not be readily apparent from these common forms. The term used to describe general shape of a crystal is ***habit***.

Some common crystal habits are as follows:

Individual Crystals

- Cubic - cube shapes.
- Octahedral - shaped like octahedrons.
- Tabular - rectangular shapes.

- Equant - minerals that have all of their boundaries of approximately equal length.
- Acicular - long, slender crystals.
- Prismatic - abundance of prism faces.
- Bladed - like a wedge or knife blade.

Groups of Distinct Crystals

- Dendritic - tree-like growths.
- Reticulated - lattice-like groups of slender crystals.
- Radiated - radiating groups of crystals.
- Fibrous - elongated clusters of fibers.
- Botryoidal - smooth bulbous or globular shapes.
- Globular - radiating individual crystals that form spherical groups.
- Drusy - small crystals that cover a surface.
- Stellated - radiating individuals that form a star-like shape.

Some minerals characteristically show one or more of these habits, so habit can sometimes be a powerful diagnostic tool.

4.1.2 Cleavage, Parting

Crystals often contain planes of atoms along which the bonding between the atoms is weaker than along other planes. In such a case, if the mineral is struck with a hard object, it will tend to break along these planes. This property of breaking along specific planes is termed cleavage. Because cleavage occurs along planes in the crystal lattice, it can be described in the same manner that crystal forms are described. For example if a mineral has cleavage along {100} it will break easily along planes parallel to the (100) crystal face, and any other planes that are related to it by symmetry. Thus, if the mineral belongs to the tetragonal crystal system it should also cleave along faces parallel to (010), because (100) and (010) are symmetrically related by the 4-fold rotation axis. The mineral will be said to have two directions of cleavage

The cleavage can also be described in terms of its quality, i.e., if the mineral cleaves along perfect planes it is said to be perfect, and if it cleaves along poorly defined planes it is said to be poor. Cleavage can also be described by general forms names, for example if the mineral breaks into rectangular shaped pieces it is said to have cubic cleavage (3 cleavage directions), if it breaks into prismatic shapes, it is said to have prismatic cleavage (2 cleavage directions), or if it breaks along basal pinacoids (1 cleavage direction) it is said to have pinacoidal cleavage.

Parting is also a plane of weakness in the crystal structure, but it is along planes that are weakened by some applied force. It therefore may not be apparent in all specimens of the same mineral, but may appear if the mineral has been subjected to the right stress conditions.

4.1.3 Fracture

If the mineral contains no planes of weakness, it will break along random directions called fracture. Several different kinds of fracture patterns are observed.

- Conchoidal fracture - breaks along smooth curved surfaces.
- Fibrous and splintery - similar to the way wood breaks.
- Hackly - jagged fractures with sharp edges.
- Uneven or Irregular - rough irregular surfaces.

4.1.4 Hardness

Hardness is determined by scratching the mineral with a mineral or substance of known hardness. Hardness is a relative scale, thus to determine a mineral's hardness, you must determine that a substance with hardness greater than the mineral does indeed scratch the unknown mineral, and similarly the unknown mineral scratches a known mineral of lesser hardness.

Hardness is determined on the basis of Moh's relative scale of hardness exhibited by some common minerals. These minerals are listed below, along with the hardness of some common objects.

Hardness	Mineral	Common Objects
1	Talc	Finger nail
2	Gypsum	Finger nail (2+)
3	Calcite	Finger nail Copper Penny (3+)
4	Fluorite	Can't scratched by nail
5	Apatite	Steel knife blade (5+), Window glass (5.5)
6	Orthoclase	Steel file
7	Quartz	Can't scratched by knife
8	Topaz	
9	Corundum	
10	Diamond	

Several precautions are necessary for performing the Hardness test:

- If you attempt to scratch a soft mineral on the surface of a harder mineral some of the softer substance may leave a mark of fine powder on the harder mineral. This should not be mistaken for a scratch on the harder mineral. A powder will easily rub off, but a scratch will occur as a permanent indentation on the scratched mineral.
- Some minerals have surfaces that are altered to a different substance that may be softer than the original mineral. A scratch in this softer alteration product will not reflect the true hardness of the mineral. Therefore it is advised that always use a fresh surface to perform the hardness test.
- Sometimes the habit of the mineral will make a difference. For example aggregates of minerals may break apart leaving the impression that the mineral is soft. Or, minerals that show fibrous or splintery habit may break easily into fibers or splinters.

In some minerals hardness is dependent on direction, it is considered as diagnostic property. For example: Kyanite has a hardness of 5 parallel to the length of the crystal, and a hardness of 7 along the direction perpendicular to the length. Similarly, calcite has a hardness of 3 for all surfaces but on basal plane {0001}it shows hardness 2.

4.1.5 Tenacity

Tenacity is the resistance of a mineral to breaking, crushing, or bending. Tenacity can be described by the following terms.

- Brittle - Breaks or powders easily.
- Malleable - can be hammered into thin sheets.
- Sectile - can be cut into thin shavings with a knife.
- Ductile - bends easily and does not return to its original shape.
- Flexible - bends somewhat and does not return to its original shape.

4.1.6 Density (Specific gravity)

Density refers to the mass per unit volume. Specific Gravity is the relative density, (weight of substance divided by the weight of an equal volume of water). In cgs units density is grams per cm^3, and since water has a density of 1 g/cm^3, specific gravity would have the same numerical value has density, but no units (units would cancel). Specific gravity is often a very diagnostic property for those minerals that have high specific gravities. In general, if a mineral has higher atomic number cations it has a higher specific gravity. For example, in the carbonate minerals the following is observed:

Mineral	Composition	Atomic # of Cation	Specific Gravity
Aragonite	$CaCO_3$	40.08	2.94
Strontianite	$SrCO_3$	87.82	3.78
Witherite	$BaCO_3$	137.34	4.31
Cerussite	$PbCO_3$	207.19	6.58

Specific gravity can usually be qualitatively measured by the weight of a mineral, in other words those with high specific gravities usually feel heavier. Most common silicate minerals have a specific gravity between about 2.5 and 3.0. These would feel light compared to minerals with high specific gravities.

For comparison, examine the following table:

Mineral	Composition	Specific Gravity
Graphite	C	2.23
Quartz	SiO_2	2.65
Feldspars	$(K,Na)AlSi_3O_8$	2.6-2.75
Fluorite	CaF_2	3.18
Topaz	$Al_2SiO_4(F,OH)_2$	3.53
Corundum	Al_2O_3	4.02
Barite	$BaSO_4$	4.45
Pyrite	FeS_2	5.02
Galena	PbS	7.5
Cinnabar	HgS	8.1
Copper	Cu	8.9
Silver	Ag	10.5

4.1.7 Colour

Colour is sometimes an extremely diagnostic property of a mineral, for example olivine and epidote are almost always green in color. But, for some minerals it is not at all diagnostic because minerals can take on a variety of colors. These minerals are said to be allochromatic. For example quartz can be clear, white, black, pink, blue, or purple.

4.1.8 Streak

Streak is the colour of powder produced by scratching the surface of mineral on streak plate. It is important diagnostic properties of mineral. Ex. Hematite has cherry red streak.

4.1.9 Lustre

Lustre refers to the general appearance of a mineral surface to reflected light. Two general types of lustre are designated as follows:

1. **Metallic** - looks shiny like a metal. Usually opaque and gives black or dark colored streak.
2. **Non-metallic** - Non metallic lustres are referred to as
   a) **Vitreous** - looks glassy - examples: clear quartz, tourmaline
   b) **Resinous** - looks resinous - examples: sphalerite, sulfur.
   c) **Pearly** - iridescent pearl-like - example: apophyllite.
   d) **Greasy** - appears to be covered with a thin layer of oil - example: nepheline.
   e) **Silky** - looks fibrous.-examples-some gypsum, serpentine, malachite.
   f) **Adamantine** - brilliant luster like diamond.

4.1.10 Play of colours

Interference of light reflected from the surface or from within a mineral may cause the color of the mineral to change as the angle of incident light changes. This sometimes gives the mineral an iridescent quality. Minerals that show this include: bornite (Cu_5FeS_4), hematite (Fe_2O_3), sphalerite (ZnS), and some specimens of labradorite (plagioclase).

4.1.11 Fluorescence and Phosphorescence

Minerals that light up when exposed to ultraviolet light, X-rays, or cathode rays are called fluorescent. If the emission of light continues after the light is cut off, they are said to be phosphorescent. Some specimens of the same mineral show fluorescence while other do not. For example some crystals of fluorite (CaF_2) show fluorescence and others do not. Other minerals show fluorescence frequently, but not always. These include - scheelite ($CaWO_4$), willemite (Zn_2SiO_4), calcite ($CaCO_3$), scapolite ($3NaAlSi_3O_8$.(NaCl - $CaCO_3$), and diamond (C).

4.1.12 Magnetism

Magnetic minerals result from properties that are specific to a number of elements. Minerals that do not have these elements, and thus have no magnetism are called **diamagnetic**. Examples of diamagnetic minerals are quartz, plagioclase, calcite, and apatite. Elements like Ti, Cr, V, Mn, Fe, Co, Ni, and Cu can sometimes result in magnetism. Minerals that contain these elements may be weakly magnetic and can be separated from each other by their various degrees of magnetic susceptibility. These are called **paramagnetic** minerals. Paramagnetic minerals only show magnetic properties when subjected to an external magnetic field. When the magnetic field is removed, the minerals have no magnetism.

4.2 Optical Properties of Rock Forming Minerals

4.2.1 Rock forming minerals

Minerals that form common rocks on the earths crust are known as rock forming minerals, these minerals are mostly silicate minerals and they are easily identified by their specific optical properties in thin section under the microscope. The microscope used to study the rocks/mineral is known as petrological microscope. Function of Petrological microscope is different from biological microscope, and it can be understood by principles of optics.

4.2.2 Principles of optics

In physics the light is considered as electro-magnetic vibration. It consists of many rays ranging from wave length (380nm to 770nm). There are two types of light:

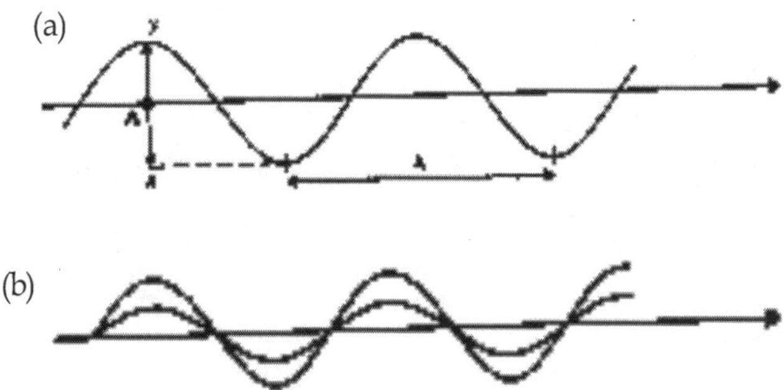

Fig. 4.1 (a) Monochromatic light of wavelength λ
(b) Two waves of the same wave length but of different amplitude.

a) Monochromatic light - Light of single wave length. It is also known as plane polarized light.

b) Normal light - In normal light the particles vibrate in all direction perpendicular to the direction of propagation.

Biological microscope is used with normal light while petrological microscope is used under polarized light/cross polarized light.

4.2.3 Petrological microscope

Petrological microscope is specially designed to study rocks and minerals in thin section. It contains an eyepiece at the top with cross-wires and objective lens at the bottom. There are two slots above objective lens for accessory plates. (Mica plate, Gypsum plate). Microscope stage is present just below objective. There is a light source (concave/flat mirror for reflecting light). Rock/mineral slide is placed at the centre of microscopic stage and optical properties exhibit under polarized and cross nicol conditions.

I. Plane polarized light and crossed nicols (Polarizer & Analyzer)

   Minerals exhibit different optical properties under plane polarized light (ppl) and crossed nicol conditions. There are two polarizing attachments, one below microscopic stage is called Polarizer and one above the microscopic stage is called Analyzer are attached.

   A. Plane Polarized light: When ordinary light reflects from mirror below passing through polarizer. It forms plane polarized light.

   B. Crossed Nicol / Crossed Polarized light: The analyzer, similar to polarizer is placed above microscope stage and oriented at right angle to polarizer. When analyzer plate inserted, it is called crossed polar condition or cross nicol condition.

II. Other accessories

   A. Bertrand lens

   It is placed above analyzer slot and used to identify interference figure.

4.2.4 Isotropic and Anisotropic minerals

Isotropic minerals are those that transmit light with equal velocity in all directions. Isotropic mineral does not show any property under cross nicol condition, it becomes dark. Example: Minerals crystallizing in cubic system or Glass. Anisotropic minerals are those that transmit light with different velocities in different directions. Optical properties are used to identify different minerals in plane polarized and crossed nicol condition.

4.2.5 Extinction angle

A) Under plane Polarized light

a) Colour	Mineral shows variety of colours Quartz & felspar are colourless.
b) Pleocroism	Under PPL when we rotate the micscope stage colour of mineral changes. This is know pleochroism.
c) From/habbit	Euhedral, Subhedral, Anhedral.
d) Relief	Relief is the sharpness of vision of mineral particularly at the boundary. Relief depends on difference between refractive index & of mineral and refractive index of enclosing cement.
e) Cleavage	Tendency of mineral to split along definite planes.
f) Alteration	Mineral when altered it appears cloudy.

B) Under crossed Nicol condition

a) Isotropic/	Minerals belonging to the cubic system are isotropic.
Anisotropic	They are dark under crossed Nicol condition. Anisotropic minerals show extinction and interference colour.
b) Interference colour	Depending of thickness of slide & orientation of minerals anisotropic mineral shows different color on rotation. Ex.Quartz shows gray.
c) Extinction	Straight or inclined.
d) Twinning	Present or absent.
e) Zoning	Present or absent.

Isotropic minerals do not show extinction angle while anisotropic minerals exhibit extinction (dark) four times during complete rotation of microscope stage under cross nicol condition. (Analiser in condition). Extinction angle may be straight/parallel or inclined and it is the diagnostic property of some minerals. The extinction angle also depends on the nature of section like basal prismatic etc. Some minerals like quartz show wavy extinction or undulose extinction which is due to deformation.

Twinning - is seen when the same mineral show two or more characters under microscope. Microcline shows cross hatched twinning, Plagioclase laths shows polysynthetic or lamellar twinning.

4.2.6 Zoning

Variation within single mineral is seen due to compositional difference in plane polarized light or cross polarized condition, it shows change in birefringence, extinction orientation.

4.3 Classification of Rock Forming Silicate Minerals

1. Neso-silicate - A group of silicate minerals having discrete (SiO_4) tetrahedra.

Neso-silicate minerals are:

i) Olivine
ii) Zircon
iii) Sphelerite
iv) Garnet
v) Andalusite
vi) Kyanite
vii) Sillimanite
viii) Topaz
ix) Staurolite

2. Sorosilicates

i) Epidote
ii) Zoisite
iii) Hemimorphite

3. Cyclosilicates

i) Beryl
ii) Tourmaline

4. Inosilicate (Chain Silicate)

Single chain **Pyroxenes**

i) Diopside ii) Augite

iii) Aegirin

Double chain **Amphibole**

i) Tremolite ii) Actinolite

iii) Hornblende iv) Asbenstos-Fibrous Amphebole

5. Phyllosilicate (Sheet)

i) Mica Muscovite , Biotite, Phlogopite. Other minerals are

• Phlogopite • Glauconite • Pyrophyllite • Talc • Chlorite • Serpentine • Keolinite • Illite • Montmorillonite • Vermiculite • Appophylite

6. Tectosilicate

(A) Felspar

1. Orthoclase-Microcline Plagioclase

(B) Quartz group

Quartz Zeolite

Oxides	Chloride	Diamond
Magnetite	Halite	Carbonate
Hematite	Fluoride	Calcite
Corundum	Fluorite	Dolomite
Ilmenile $FeTiO_3$	Sulphide	Sulphate
Pyrolusile MnO_2	Galena	Gypsum
Cassiterite SnO_2	Realgar	Phosphate
Psilomelane	Orpiment	Apatite
Bauxite	Pyrite	
Goethite		
Limonite		

4.4 Description of Megascopic Minerals (Physical and Optical Properties)

4.4.1 Olivine

Olivine group of minerals consist of isomorphos series (R_2SiO_4) in which R = Mg^{+2} or Fe^{+2} Mg_2SiO_4 forsterite Fe_2SiO_4 Fayalite.

Physical Properties

1. Colour - Shades of green, pale green, olive green, yellow - forsterite (black) fayalite
2. Form (Habit) - Prismatic, Massive, compact
3. Cleavage - Poor, one directional
4. Fracture - Present
5. Lustre - Vitreous
6. Hardness - 6-7
7. Sp. Gravity - 3.22 (Forsterite) 4.39 (Fayalite)
8. Crystal system - Orthorhombic
9. Diagnostic prop - Colour, Fracture.
10. Name - Olivine

Optical properties

(A) Under plane Polarized light

1. Isotopic / Anisotopic - Anisotopic
2. Colour - Colourless or faint green
3. Form/ Habit - Euhedral in extrusive rock, Anhedral in plutonic rock
4. Cleavage - One directional
5. Relief - High to very high
6. Alteration - Olivine generally alters to serpentine

(i) Under crossed Nicol condition

7. Birefringence - High (Third order interference colours)
8. Extinction - Parallel/ Straight
9. Name - Olivine

4.4.2 Zircon $ZrSiO_4$

Physical Properties

1.	Colour	-	Grey, yellow, greenish, reddish brown
2.	Form/Habit	-	Prismatic
3.	Cleavage	-	Present two directional
4.	Fracture	-	Concholidal
5.	Lustre	-	Admantine
6.	Handness	-	7.5
7.	Sp. Gravity	-	4.6-7
8.	Crystal system	-	Tetragonal
9.	Diagnostic	-	Crystal system, hardness, luster
10.	Name	-	Zircon

Optical properties

A. Under plane Polarized

1.	Colour	-	Colourless
2.	Form/Habit	-	Prismatic
3.	Cleavage	-	One direction perfect
4.	Relief	-	High
5.	Alteration	-	High to very high
6.	Pleochroism	-	No.

Under crossed Nicol Light

1.	Isotropic/ Anisotropic	-	Basal Sections Isotropic Prismatic Sections Anisotropic
2.	Interference colour	-	• Prismatic sections 3rd order colours
	Birefringence	-	• Higher order like Olivine
3.	Extinction	-	• Parallel
4.	Diagnostic property	-	Form, Relief, Birefringence
5.	Name	-	• Zircon

4.4.3 Sphene

A. Physical properties

1. Colour - Brown, grey, green, yellow or Black
2. Form/Habit - Wedge shaped, massive
3. Cleavage - One directional
4. Fracture - Conchoidal imperfect
5. Lustre - Resinous/Admantine
6. Hardness - 5
7. Sp. Gravity - 3.45-3.55
8. Crystal system - Monoclinic
9. Diagnostic property - Crystal system
10. Name of Min - Sphene

B. Optical properties

(i) Under uncross condition/Plane Polarized light (ppl)

1. Colour - Colourless, pale yellow
2. Form/Habit - Irregular four sides
3. Cleavage - One directional good
4. Relief - Extremely high
5. Altration - Alters to leucoxene.
6. Pleochroism - Pink yellowish/pink to orange

(ii) Under crossed Nicol condition

1. Isotropic/ Anisotropic - Anisotropic
2. Interference Birefringence - Present Anomolous
3. Extinction - Inclined
4. Twinning - Twinning present
5. Diagnostic property - Form, Relief, Extinction
6. Name - Sphene

4.4.4 Garnet

A. Physical properties

1. Colour - Olive green, pink, crimson, red etc.
2. Form - Dodecaheatron, Tetrahexahedron
3. Cleavage - Absent, Parting-Irregular cracks
4. Fracture - Imperfect conchoidal
5. Lustre - Vitreous
6. Hardness - 7-7.5
7. Sp. Gravity - 3.5
8. Crystal system - Cubic
9. Diagnostic property - Crystal system colour, cleavage
10. Name of mineral - Garnet

B. Optical properties

(i) Under Plane Polarized Light (ppl)

1. Colour - Colourless, pale pinkish
2. Form/Habit - Subhedral, branching cracks
3. Relief - High to very High
4. Cleavage - Absent, parting
5. Alteration - Chlorite
6. Pleochroism - Anomalous (weak)

(ii) Under crossed Nicol condition

1. Isotropic/ Anisotropic - Isotropic
2. Interference colour/Birefringenc - Dark under Crossed Nicol
3. Extinction/Angle
4. Twining - Complex
5. Diagnostic property - Cubic. No cleavage Isotropic
6. Name - Garnet

4.4.5 Andalusite

A. Physical properties

1. Colour - Pearl grey
2. Form - Columnar, square
3. Cleavage - Poor, one direction (Prismatic 110)
4. Fracture - Uneven
5. Lustre - Vitreous
6. Hardness - 6.5-7.5
7. Sp. Gravity - 3.13-3.16
8. Crystal system - Orthorhombic
9. Diagnostic Property - Form and cleavage
10. Name of Min - Andalusite

B. Optical properties

(i) Under Plane Polarized light (Uncross condition)

1. Colour - Colourless, but thick sections exhibit pleochroism
2. Form/Habit - Square, poikiloblastic inclusions
3. Relief - Medium
4. Cleavage - 2 sets at right angle
5. Alteration - Alters to sericite, Associated with Sillimanite
6. Pleochrosim - Weak, Patchy, Pink & Green

(ii) Under crossed Nicol condition

1. Isotropic/ Anisotropic - Anisotropic
2. Interference colour - Ist order gray
3. Extinction - Parallel/ Straight
4. Twining - Absent
5. Diagnostic properties - Cleavage, Form, Extinction
6. Name - Andalusite

4.4.6 Kyanite

A. Physical properties

1.	Colour	-	Light blue & white
2.	Form	-	Bladed, prismatic
3.	Streak	-	White
4.	Clearage	-	Two directional
5.	Fracture	-	Uneven
6.	Lustre	-	Pearly
7.	Hardness	-	5.5 to 7.00 (Lover along length and higher across length)
8.	Sp. Gravity	-	3.58 to 3.65
9.	Crystal system	-	Triclinic
10.	Diagnostic property	-	Form, colour, hardness
11.	Name	-	Kyanite

B. Optical properties

(i) Under Plane Polarized light (ppl)

1.	Colour	-	Colourless, pale blue
2.	Form	-	Subhedral, bladed, Prismatic
3.	Relief	-	Very high
4.	Cleavage	-	Two directional
5.	Alteration	-	Alters to Serecite
6.	Pleochroism	-	Weak

(ii) Under crossed Nicol condition

1.	Isotropic/ Anisotropic	-	Anisotropic
2.	Birefringence	-	Low
3.	Extinction	-	Straight, Parallel
4.	Twining	-	Polysynthetic twinning
5.	Extinction	-	Inclined 0° to 32°
	Diagnostic Property	-	Form, relief, extinction
6.	Name of Mineral	-	Kyanite

4.4.7 Sillimanite

A. Physical properties

1. Colour - Shades of brown, grey, green
2. Form - Acicular crystal, diamond shape
3. Cleavage - One directional good
4. Fracture - Uneven
5. Lustre - Vitreous
6. Hardness - 6.5-7.5
7. Sp. Gravity - 3.23-3.27
8. Crystal system - Orthorhombic
9. Diagnostic property - Form, Hardness
10. Origin - Metamorphic mineral
11. Name of Mineral - Sillimanite

B. Optical properties

(i) Under Plane Polarized light (ppl)

1. Colour - Colourless
2. Form - Needle like, slender prism
3. Relief - High
4. Pleochroism - Weak
5. Cleavage - Two sets
6. Alteration - Alters to sericite

(ii) Under crossed Nicol condition

1. Isotropic/ Anisotropic - Anisotropic
2. Interference Colour - 1st order red
3. Extinction - Parallel
4. Twining - Absent
5. Diagnostic Property - Form, cleavage, extinction.
6. Name of Mineral - Sillimanite

Cyclosilicates

4.4.8 Tourmaline

A. Physical Properties

1.	Colour	-	Usually pitch black
2.	Form/habit	-	Crystalline
3.	Cleavage	-	Imperfect
4.	Fracture	-	Uneven
5.	Lustre	-	Vitreous, shining
6.	Hardness	-	7.0-7.5
7.	Sp. Gravity	-	2.98-3.20
8.	Crystal system	-	Trigonal
9.	Diagnostic properties	-	Black colour with crystalline form hardness, striated surface
10.	Name of mineral	-	Tourmaline

B. Optical Properties

(i) Under Plane Polarized light

1.	Colour	-	Dark brown, green, yellow etc
2.	Form/habit	-	Usually sub-hedral, prismatic
3.	Cleavage	-	Imperfect
4.	Relief	-	High
5.	Pleochroism	-	Pleochroic
6.	Alteration	-	Absent

(ii) Under Crossed Nicol condition

1.	Isotropic/ Anisotropic	-	Anisotropic
2.	Interference colour	-	Second order green, brown, yellow etc
3.	Extinction	-	Straight
5.	Twining	-	Absent
6.	Diagnostic properties	-	Dark brown to green pleochroic with straight extinction and high relief
7.	Name of mineral	-	Tourmaline

4.4.9 Beryl

C. Physical Properties

1. Colour - Emerald green, pale green or brown (streak-uncolured)
2. Form/habit - Crystalline, massive
3. Cleavage - 1 set basal
4. Fracture - Conchoidal or uneven
5. Lustre - Vitreous
6. Hardness - 7.5 - 8
7. Specific gravity - 3.5 - 3.6
8. Crystal system - Hexagonal
9. Diagnostic properties - Green, palegreen, crystalline form with great hardness (7.5 to 8)
10. Name of mineral - Beryl

D. Optical Properties

(iii) Under uncross condition/Plane Polarized light

1. Colour - Colourless
2. Form/habit - Prismatic crystals
3. Cleavage - Basal cleavage one set/poor
4. Relief - Low to moderate
5. Pleochroism - Non-pleochroic
6. Alteration - Easily alteres to clay mineral

(iv) Under crossed Nicol Condition

1. Isotropic/ Anisotropic - Anisotropic
2. Interference Colour - First order gray
3. Extinction - Straight
4. Twining - Absent
5. Diagnostic Properties - Low relief, straight extinction
6. Name of Mineral - Beryl

Inosilicate - Chain silicate

4.4.10 Augite (Pyroxene)

A. Physical Properties

1. Colour - Black and greenish black
2. Form/habit - Also occur as massive, crystalline
3. Cleavage - Perfect 2 set (at 90 degree)
4. Fracture - Uneven
5. Lustre - Vitreous
6. Hardness - 5-6
7. Sp. Gravity - 3.2-3.5
8. Crystal system Monoclinic
9. Diagnostic Properties - Black colour massive form with 2 set cleavage (at 90 degree)
10. Name of mineral - Augite

B. Optical Properties

(i) Under ppl

1. Colour - Colourless
2. Form/habit - Usually sub-prismatic
3. Cleavage - Perfect 2 set
4. Relief - High
5. Pleochroism - Non-pleochroic
6. Altration - Sometimes alters to hornblende

(ii) Under crossed Nicol Conditions

1. Isotropic/ Anisotropic - Anisotropic
2. Interference Colour - Second order brown, yellow etc.
3. Extinction - Inclined
4. Extinction Angle - 45°-54°
5. Twinning - Commonly polysynthetic
6. Diagnostic Properties - Colourless, 2 sets cleavage and Inclined extinction (45° to 54°)
7. Name of Mineral - Augite

4.4.11 Diopsides

C. Physical Properties

1. Colour - Colourless, white greenish
2. Form/habit - Short prismatic crystals, granular
3. Cleavage - Perfect 2 sets basal (87°)
4. Fracure - Even
5. Lustre - Vitreous
6. Hardness - 5.5 to 6.5
7. Sp. Gravity - 3.22 to 3.56
8. Crystal system - Monoclinic
9. Diagnostic Properties - Colourless 2 sets basal cleavage with hardness 5.5 to 6.5
10. Name of Mineral - Diopside

D. Optical Properties

(i) Under uncross condition/Plane Polarized light

1. Colour - Colourless/Gray/ Green
2. Form/habit - Usually sub-prismatic
3. Cleavage - Perfect 2 set
4. Relief - High
5. Pleochroism - Non-pleochroic/ Weakly pleochroic
6. Altration - Often alters to tremolite, actinolite

(ii) Under crossed Nicol Conditions

1. Isotropic/ Anisotropic - Anisotropic
2. Interference Colour - Second order green brown, yellow etc
3. Extinction - Inclined
4. Twinning - Polysynthetic
5. Diagnostic properties - Colourless, 2 set clearage, polysynthetic twinning
6. Name of Mineral - Diopside

Inosilicates - Double chain (Amphibole)

4.4.12 Asbestos

Mineralogically, fibrous forms of Amphebole (tremolite/actinolite) are called Asbestos. But commercially, term Asbestos includes fibrous Anthophyllite, Crosidolite, Crystalite (fibrous serpentine) etc.

A. Physical properties

1. Colour - White, greenish, Brownish
2. Form/habit - Fibrous
3. Cleavage - One directional
4. Fracture - Uneven
5. Lustre - Pearly/Silky
6. Hardness - 5 to 6
7. Sp. Gravity - Medium
8. Crystal system - Monoclinic
9. Diagnostic properties - Fibrous form
10. Name of mineral - Asbestos (Actinolite, Tremolite)

B. Optical Properties

(i) Under Plane Polarized light

1. Colour - Colourless (Tremolite)/Brownish green (Actinolite)
2. Form/habit - Prismatic
3. Cleavage - One directional perfect
4. Relief - Moderate to high
5. Pleochroism - Pale yellow to greenish blue
6. Alteration - Alters to chlorite

(ii) Under crossed Nicol Conditions

1. Isotropic/ Anisotropic - Anisotropic
2. Interference Colour - Second order green
3. Extinction - Inclined
4. Twinning -
5. Diagnostic properties - Form Clearage
6. Name of Mineral - Asbestos (Actinolite, Tremolite)

4.4.13 Hornblende

C. Physical Properties

1.	Colour	-	Black or greenish black (Streak-uncoloured)
2.	Form/habit	-	Also occur as blade like prism or massive
3.	Cleavage	-	Perfect 2 set basal (Oblique at 56°-124°)
4.	Fracture	-	Uneven
5.	Lustre	-	Vitreous
6.	Hardness	-	5 to 6
7.	Sp. Gravity	-	3.3 - 3.7
8.	Crystal system		Monoclinic
9.	Diagnostic Properties	-	Greenish black with two set cleavage and 5 to hardness
10.	Name of Mineral	-	Hornblende

D. Optical Properties

(iii) Under uncross condition/Plane Polarized light

1.	Colour	-	Deep green to yellow
2.	Form/habit	-	Usually sub-hedral prismatic
3.	Cleavage	-	Perfect 2 sets (56 to 120 degree)
4.	Relief	-	High
5.	Pleochroism	-	Pleochroic
6.	Altration	-	Often alters to chlorite

(iv) Under crossed Nicol Conditions

1.	Isotropic/ Anisotropic	-	Anisotropic
2.	Interference Colour	-	Second order green brown, yellow etc
3.	Extinction	-	Inclined extinction
4.	Extinction Angle	-	12°-30°
5.	Twinning	-	Simple / Polysynthetic
6.	Diagnostic Properties	-	Deep green colour with inclined extinction with 12°-30°
7.	Name of Mineral	-	Hornblende

Double Chain

4.4.14 Tremolite

A. Physical Properties

1. Colour - Colourless, brownish, green
2. Form/habit - Elongate, prismatic, acicular
3. Cleavage - One direction perfect
4. Fracture - Uneven
5. Lustre - Vitreous
6. Hardness - 5-6
7. Sp. Gravity - 3.02-3.44
8. Crystal system - Monoclinic
9. Diagnostic properties - Form cleavage
10. Name of mineral - Tremolite

B. Optical Properties

(i) Under Plane Polarized light

1. Colour - Colourless brown green
2. Form/habit - Elongate, Accicular
3. Cleavage - 2 sets
4. Relief - Moderate to high
5. Pleochroism - Medium pleochroic
6. Alteration - Alters to chlorite

(ii) Under crossed Nicol condition

1. Isotropic/ Anisotropic - Anisotropic
2. Interference Colour - IInd order green.
3. Extinction - Inclined 21°
4. Twinning - Present
5. Diagnostic Properties - Interference colour, form. inclined extinction
6. Name of Mineral - Tremolite

4.4.15 Actinolite

C. Physical Properties

1. Colour - Brownish Green
2. Form/habit - Fibrous, Prismatic elongate
3. Cleavage - Perfect one directional
4. Fracture - Uneven
5. Lustre - Vitreous
6. Hardness - 5 to 6
7. Sp. Gravity - 3.2 to 3.44
8. Crystal system - Monoclinic
9. Diagnostic Properties - Form, cleavage
10. Name of Mineral - Actinolite

D. Optical Properties

(iii) Under Plane Polarized Light

1. Colour - Brownish Green
2. Form/habit - Fibrous, prismatic
3. Cleavage - One directional perfect (prismatic)
4. Relief - Moderate to high
5. Pleochroism - Pale yellow to yellowish green
6. Alteration - Alters to chlorite

(iv) Under crossed Nicol condition

1. Isotropic/Anisotropic - Anisotropic
2. Interference colour - IInd order green
3. Extinction - Inclined
4. Extinction Angle - 11°
5. Twinning - Present
6. Diagnostic properties - Form, interference colour and extinction
7. Name of Mineral - Actinolite

4.4.16 Phyllosilicates (Sheet structure)

A. Muscovite

1. Colour - Colourless, white, (streak-uncoloured)
2. Form/habit - Usually sub-hedral, tabular, flaky, enystalline
3. Cleavage - Perfect one set basal
4. Fracture - Very difficult to obtain due to perfect clearage
5. Lustre - Pearly, transparent
6. Hardness - 2-2.5
7. Sp. Gravity - 2.76-3.0 light in weight
8. Crystal system - Monoclinic
9. Diagnostic Properties - Colourless micaceous form with basal cleavage and low hardness
10. Name of mineral - Muscovite

B. Optical Properties

(i) Under Plane Polarized light

1. Colour - Colourless
2. Form/habit - Usually subhedral
3. Cleavage - Perfect one set
4. Relief - High, sometimes show faint twinkling
5. Pleochroism - Non-pleochroic
6. Altration - Absent

(ii) Under crossed Nicol condition

1. Isotropic/ Anisotropic - Anisotropic
2. Interference Colour - Higher order colour
3. Extinction - Straight
4. Twinning - Absent
5. Diagnostic Properties - Colourless, one set cleavage & higher relief and straight extinction
6. Name of Mineral - Muscovite

4.4.17 Biotite

A. Physical Properties

1.	Colour	-	Black or dark brown (Streak-neutral)
2.	Form/habit	-	Crystalline, prismatic, tabular flaky
3.	Cleavage	-	Perfect basal, one set
4.	Fracture	-	Very difficult to obtain
5.	Lustre	-	Splendant, pearly
6.	Hardness	-	2.5 to 3.0
7.	Sp. Gravity	-	2.7 to 3.1
8.	Crystal system	-	Monoclinic
9.	Diagnostic Properties	-	Dark coloured micaceous form one set basal cleavage and 2.5 to 3.0 hardness (low)
10.	Name of Mineral	-	Biotite

B. Optical Properties

(iii) Under Plane Polarized light

1.	Colour	-	Reddish brown to yellow, strongly colored
2.	Form/habit	-	Usually euhedral, tabular
3.	Cleavage	-	Perfect, one set
4.	Relief	-	High
5.	Pleochroism	-	Pleochroic (Maximum absorption parallel to polariser)
6.	Altration	-	Often alter to chlorite

(iv) Under crossed Nicol condition

1.	Isotropic/ Anisotropic	-	Anisotropic
2.	Interference Colour	-	Higher order colour
3.	Extinction	-	Straight extinction
4.	Twinning	-	Absent
5.	Diagnostic properties	-	Reddish brown to yellow, colour with perfect one set cleavage and straight extinction
6.	Name of Mineral	-	Biotite

Tectosilicate

4.4.18 Plagioclase

A. Physical Properties

1. Colour - Dark grey to bluish
2. Form/habit - Prismatic
3. Cleavage - Two sets (Prismatic/Basal)
4. Fracture - Conchoidal
5. Lustre - Vitreous to sub vitreous, showing play of colours.
6. Hardness - 6
7. Sp. Gravity - 2.63
8. Crystal system - Triclinic
9. Diagnostic Properties - Colour, clearage, Hardness
10. Name of mineral - Plagioclase

B. Optical Properties

(i) Under uncross condition/Plane Polarized light:

1. Colour - Colourless, cloudy, due to alteration to clay and other minerals
2. Form/habit - Prismatic
3. Cleavage - Two sets at 90°
4. Relief - Low
5. Pleochroism - Absent
6. Alteration - Alters to clay (sericite/kaolinite)

(ii) Under crossed Nicol condition

1. Isotropic/ Anisotropic - Anisotropic
2. Interference Colour - Ist order colour is darker as Ca content increases.
3. Extinction - Inclined
4. Twinning - Polysynthetic lamellar twinning

5. Diagnostic properties - Inclined Extinction, Polysynthetic twin ing, complex Albite, Carlsbad laws, sometimes pericline Bavino twinning.
6. Name of Mineral - Plagioclase

4.4.19 Quartz

C. Physical Properties

1. Colour - Colourless, white, varieties are common
2. Form/habit - Massive or crystlline
3. Cleavage - Absent/streak-absent
4. Fracture - Conchoidal
5. Lustre - Vitreous occasional resinous
6. Hardness - 7
7. Sp. Gravity - 2.65
8. Crystal system - Hexagonal
9. Diagnostic Properties - Cleavage, Hardness
10. Name of Mineral - Quartz

D. Optical Properties

(iii) Under uncross condition/Plane Polarized light

1. Colour - Colourless
2. Form/habit - Anhedral in porphyritic Igneous Rock
3. Cleavage - Absent
4. Relief - V. Low
5. Pleochroism - Absent
6. Alteration - No

(iv) Under crossed Nicol Conditions

1. Isotropic/Anisotropic - Anisotropic
2. Interference Colour - 1st order grey
3. Extinction - Wavy
4. Twinning - Not seen in thin section

5. Diagnostic Properties - Wavy extinction, cleavage absent
6. Name of Mineral - Quartz

4.4.20 Zeolite

A. Physical Properties

1. Colour - Colorless, white
2. Form/habit - Tabular/Prismatic/Fibrous/Radiating
3. Cleavage - One directional
4. Fracture - Sub-conchoidal
5. Lustre - Vitreous
6. Hardness - 5.5
7. Sp. Gravity - 2,24
8. Crystal system - Cubic
9. Diagnostic Properties - Colour, Form, Hardness
10. Name of mineral - Zeolite

B. Optical Properties

(i) Under uncross condition/Plane Polarized light

1. Colour - Colourless
2. Form/habit - Fibrous, Prismatic, rhombohedral
3. Cleavage - One directional perfect
4. Relief - Low
5. Pleochroism - Low
6. Alteration - Zeolite are hydrous silicates, formed by alteration.

(ii) Under crossed Nicol Conditions

1. Isotropic/ Anisotropic - Anisotropic
2. Interference Colour - Weak, Ist order colours
3. Extinction - Straight
   Twinning - Inter-penetrative
4. Diagnostic Properties - Twinning, Extinction
5. Name of Mineral - Zeolite

Tectosilicates

4.4.21 Orthoclase

A. Physical Properties

1. Colour - White, flesh colourd
2. Form/habit - Cyrstalline, prismatic
3. Cleavage - Perfect 2 set nearly at right angle
4. Fracture - Uneven, sometomes conchoidal
5. Lustre - Vitreous to pearly
6. Hardness - 6-7
7. Sp. Gravity - 2.57
8. Crystal system - Monoclinic
9. Diagnostic Properties - Perfect 2 sets cleavage, hardness and light specific gravity 2.57
10. Name of mineral - Orthoclase

B. Optical Properties

(i) Under Plane Polarized light

1. Colour - Colourless clouded due to alteration to clays.
2. Form/habit - Usually sub-hedral , sometimes euhedral in igneous porphyry rocks.
3. Cleavage - Perfect 2 sets
4. Relief - Low
5. Pleochroism - Non-pleochroic
6. Alteration - Alters to clay (Kaolinite)

(ii) Under crossed Nicol Conditions

1. Isotropic/ Anisotropic - Anisotropic
2. Interference Colour - 1st order grey
3. Extinction - Inclined
4. Extinction Angle - 5° to 12°
5. Twinning - Simple twinning (Calsbad law)
6. Diagnostic properties - Colourless, Inclined extinction 5° to 12°, simple twinning

7. Name of Mineral: Orthoclase

4.4.22 Microcline

A. Physical Properties

1. Colour - Grayish white, flesh red, bright green
2. Form/habit - Crystalline, prismatic
3. Cleavage - Perfect 2 sets
4. Fract - Uneven, sometimes conchoidal.
5. Lustre - Vitreous
6. Hardness - 6-6.5
7. Sp. Gravity - 2.65
8. Crystal system - Triclinic
9. Diagnostic Properties - Grayish colour, 2 sets of cleavage and hardness (6 to 6.5)
10. Name of Mineral - Microcline

B. Optical Properties

(i) Under Plane Polarized light

1. Colour - Colourless
2. Form/habit - Usually subhedral
3. Cleavage - Perfect 2 sets
4. Relief - Low
5. Pleochroism - Non-pleochroic
6. Alteration - Alters to clays (Kaolinite)

(ii) Under Crossed Nicol Conditions

1. Isotropic/Anisotropic - Anisotropic
2. Interference Colour - 1st order grey
3. Extinction - Inclined
4. Extinction Angle - 5°-15°
5. Twinning - Cross hatched twinning
6. Diagnostic Properties - Colourless, 1st order grey and cross hatched twinning

7. Name of Mineral - Microcline [Many perthites are potash felspar perthites which contain lamellae or patches of alkali plagioclase (albite, orthoclase)]

4.5 Non Metallic Minerals: Identified only on the basis of Physical Properties

4.5.1 Bauxite (Al_2O_3)

1. Colour - Reddish white, brownish white
2. Form/habit - Pisolitic, Amorphous massive
3. Streak - White
4. Lustre - Dull and earthy
5. Cleavage - Absent
6. Fracture - Uneven
7. Hardness - 2.5-3.0
8. Sp. gravity - 2.35
9. Crystal system - Amorphous
10. Diagnostic properties - Brownish white to mottled colour pisolitic structure dull lustre
11. Name of Mineral - Bauxite

4.5.2 Halite

1. Colour - Colourless to white
2. Form/habit - Crystalline, cubic, often massive
3. Streak - White
4. Lusture - Vitreous, translucent
5. Cleavage - Perfect 3 sets (cubic)
6. Fracture - Uneven
7. Hardness - 2 to 2.5
8. Sp. gravity - 2.2
9. Crystal system - Isometric (cubic)
10. Diagnostic properties- Colourless crystalline 3 set cubic cleavage very low hardness 2-2.5
11. Name of Mineral Halite

4.5.3 Fluorite (CaF_2)

1. Colour - Colourless to purple, white, green
2. Form/habit - Crystalline
3. Streak - Neutral to grey white
4. Lustre - Vitreous
5. Cleavage - Perfect 4 sets & octahedral cleavage
6. Fracture - Uneven to conchoidal
7. Hardness - 4
8. Sp. gravity - 3 to 2.25 (light in weight)
9. Crystal system - Isometric (cubic)
10. Diagnostic properties - Colourless 4 sets cleavage with hardness 4 in Mohr's scale, perfectly crystalline.
11. Name of mineral Fluorite

4.5.4 Corundum (Al_2O_3)

1. Colour - Grey, greenish or reddish, flesh colored
2. Form/habit - Crystalline or massive
3. Cleavage - Absent Basal parting
4. Fracture - Conchoidal or uneven
5. Lustre - Vitreous to dull
6. Hardness - 9 (very hard mineral)
7. Sp. Gravity - 3.52-4.1 (high sp. gravity)
8. Crystal system - Hexagonal system
9. Diagnostic Properties - Vitreous lusture with Hardness 9 and 3.52 to 4.1 (high sp. gravity)
10. Name of mineral - Corundum.

B. Optical Properties

(i) Under Plane Polarized light

1. Colour - Colourless
2. Form/habit - Rounded crystal, Tabular (Trigonal Crystal System)
3. Cleavage - Absent but basal parting prominent

4. Relief - Very High
5. Pleochroism - Non pleochroic
6. Alteration - Alters to muscovite (Metasomatic)

(ii) Under Crossed Nicol Conditions

1. Isotropic/ Anisotropic - Anisotropic
2. Interference Colour - Low, first order gray-white
3. Extinction - Parallel/ Symmetrical.
4. Twinning - Lamellar twining, colour banding
5. Diagnostic properties - High relief, Symmetrical extinction and low order interference color
6. Name of Mineral: - Corundum

4.6 Metallic Minerals (Opaque minerals)

Metallic minerals are opaque in nature. Polished sections of opaque minerals are studied in reflected light. Here only the description of physical properties is given:

Oxides

4.6.1 Pyrolusite

A. Physical Properties

1. Colour - Iron grey, or dark steel grey
2. Form/habit - Massive, pseudomorphous, reniform (Kidney shaped)
3. Cleavage - Absent
4. Fracture - Uneven or brittle
5. Lustre - Sub metallic
6. Hardness - 2 to 2.5
7. Sp. Gravity - 4.8
8. Crystal system - Orthorhombic
9. Diagnostic Properties - Iron gray colour, brittle substance and specific gravity 4.8
10. Name of mineral - Pyrolusite

4.6.2 Psilomelane

C. Physical Properties

1.	Colour	-	Iron black
2.	Form/habit	-	Amorphous, massive, botryoidal (Bunch of grapes) reniform (Kidney shaped)
3.	Cleavage	-	Absent
4.	Fracture	-	Sub-concholdal to uneven
5.	Lustre	-	Sub-metallic, dull
6.	Hardness	-	5-6
7.	Sp. Gravity	-	3.7-4.7
8.	Crystal system	-	Amorphous
9.	Diagnostic properties	-	Iron black color, botryoidal form hardness (5-6)
10.	Name of Mineral	-	Psilomelane

4.6.3 Magnetite

A. Physical Properties

1.	Colour	-	Iron black dark
2.	Form/habit	-	Crystalline to massive
3.	Cleavage	-	Imperfect
4.	Fracture	-	Sub conchoidal to regular
5.	Lustre	-	Metallic
6.	Hardness	-	5.5-6.5
7.	Sp. Gravity	-	5.18
8.	Crystal system	-	Isometric (Octahedral)
9.	Diagnostic Properties	-	Iron black colour, sp. gr. Crystalline form
10.	Name of mineral	-	Magnetite

4.6.4 Hematite

1.	Colour	-	Steel grey to iron black
2.	Form/habit	-	Massive sometimes granular & micaceous flaky (specularite)
3.	Streak	-	Cherry red

4. Cleavage - Absent
5. Fracture - Uneven to sub-conchoidal
6. Lustre - Metallic, to sub metallic, and dull
7. Hardness - 5.5 to 6.5
8. Sp. Gravity - 4.9 to 5.3
9. Crystal system - Hexagonal
10. Diagnostic Properties - Cherry red streak [sp. gravity (Heavy)]
11. Name of Mineral - Hematite

4.6.5 Ilmenite ($FeTiO_2$)

A. Physical Properties

1. Colour - Iron black
2. Form/habit - Crystalline or massive
3. Cleavage - Parting
4. Fracture - Conchoidal
5. Lustre - Sub-metallic
6. Hardness - 5 to 6
7. Sp. Gravity - 4.5 to 5.0 (Heavy)
8. Crystal system - Trigonal
9. Diagnostic Properties - Iron black crystalline nature hardness (5 to 6)
10. Name of Mineral - Ilmenite

Opaque mineral studied by ore microscope

4.6.6 Galena (PbS)

Physical Properties

1. Colour - Lead-grey
2. Form/habit - Crystalline, cubic, also massive
3. Streak - Lead-grey
4. Lustre - Metallic, splendent
5. Cleavage - Perfect 3 sets, cubic
6. Fracture - Flat, even
7. Hardness - 2.5
8. Sp. Gravity - 7.4-7.6

9.	Crystal system	-	Cubic system
10.	Diagnostic properties	-	Lead grey colour, Soft, (Low hardness) & high specific gravity (7.4-7.6)
11.	Name of Mineral	-	Galena

4.6.7 Realgar As2S2 with Arsenic 70.1%

1.	Colour	-	Fine red or orange
2.	Form/habit	-	Usually massive or granular, rarely occurs as prismatic crystal
3.	Streak	-	Red or orange
4.	Lustre	-	Resinous
5.	Cleavage	-	Perfect in crystal one set
6.	Fracture	-	Conchoidal
7.	Hardness	-	1.5-2
8.	Sp. Gravity	-	3.56,
9.	Diagnostic properties	-	Orange colour granular form, low hardness (1.5 to 2)
10.	Crystal system	-	Monoclinic system
11.	Name of Mineral	-	Realgar

4.6.8 Orpiment

1.	Colour	-	Fine lemon, yellow
2.	Form/habit	-	Crystalline, usually massive or granular, rarely crystalline
3.	Streak	-	Yellow
4.	Lustre	-	Resinous
5.	Cleavage	-	Perfect in crystal one set
6.	Fracture	-	Conchoidal
7.	Hardness	-	1.5 to 2
8.	Sp. Gravity	-	3.4 to 3.5 (Light in weight)
9.	Crystal system	-	Monoclinic system
10.	Diagnostic properties	-	Lemon yellow massive form, low hardness 1.5 to 2
11.	Name of Mineral	-	Orpiment

4.6.9 Calcite ($CaCO_3$)

Physical Properties

1. Colour - Colourless, white to various colours
2. Form/habit - Crystalline, prismatic
3. Streak - White (Neutral to whitegrey)
4. Lustre - Vitreous to pearly shining
5. Cleavage - Perfect 3 sets, rhombohedral
6. Fracture - Conchoidal, obtained with difficulty
7. Hardness - 3.0
8. Sp. gravity - 2.71 (light in weight)
9. Crystal system - Hexagonal
10. Diagnostic properties - Colourless Trigonal Class rhombohedral cleavage low hardness-3, Crystalline nature
11. Name - Calcite

4.6.10 Dolomite [$CaMg(CO_3)_2$]

1. Colour - White to grey, various colors due to compositional change
2. Form/habit - Crystalline, massive, granular
3. Streak - White
4. Lustre - Vitreous, glistening
5. Cleavage - Perfect
6. Fracture - Conchoidal or uneven
7. Hardness - 3.5-4
8. Sp. Gravity - 2.8-2.9
9. Crystal system - Hexagonal (Rhombohedral)
10. Diagnostic properties - White colored granular form, and hardness (3.5 to 4)
11. Name of Mineral - Dolomite

(See colour version on page 213

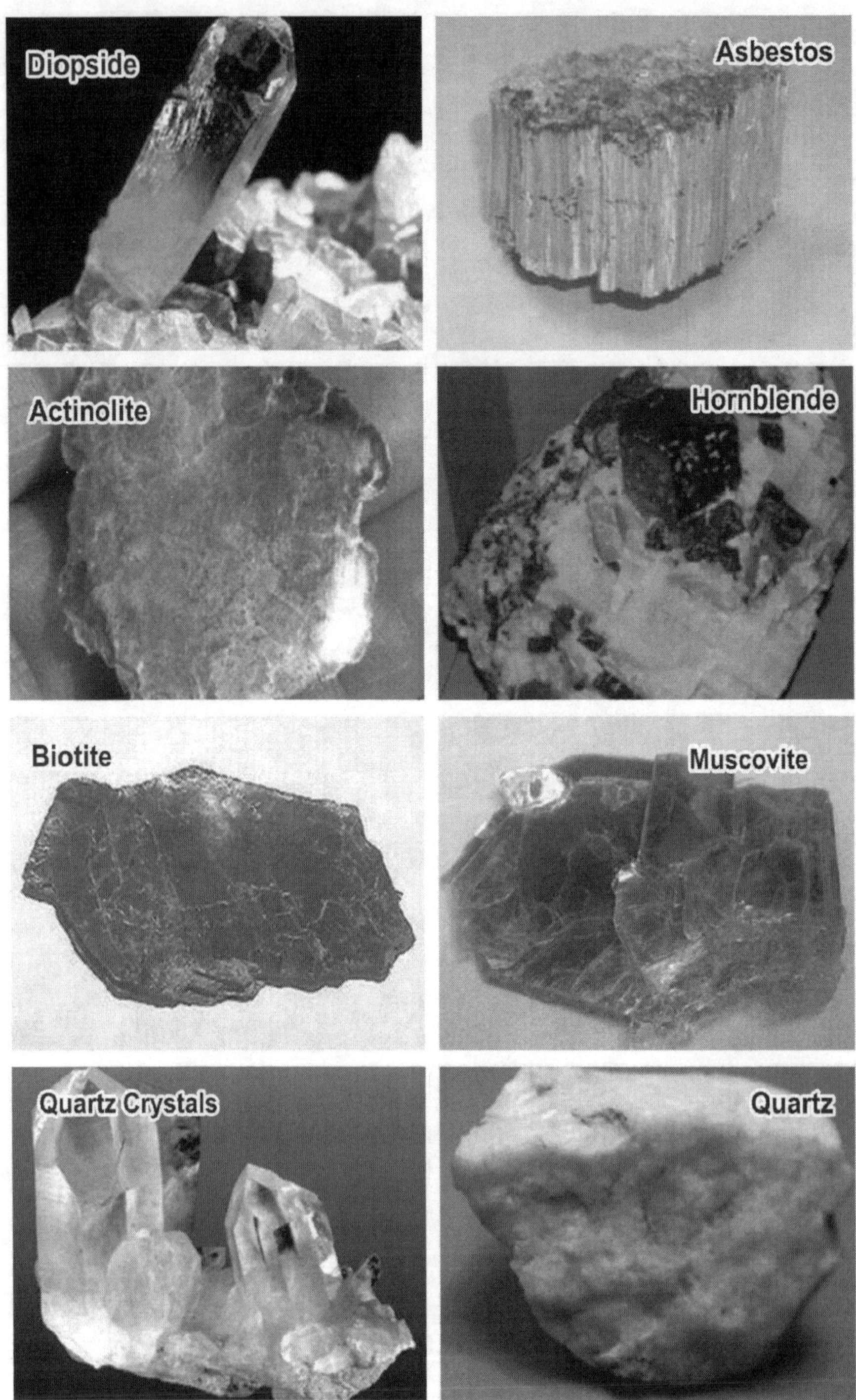

(*See colour version on page 214*)

(*See colour version on page 215*

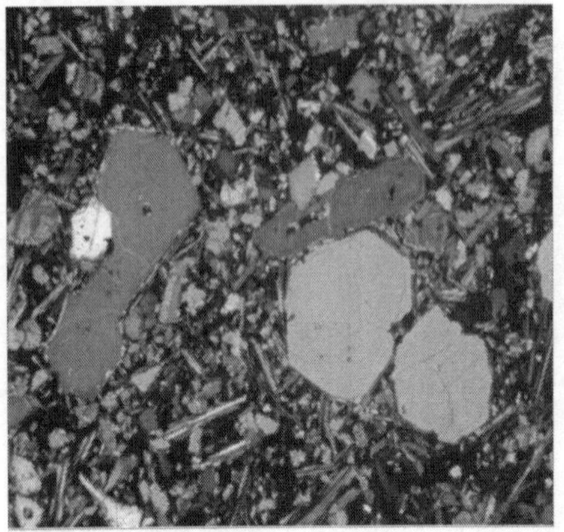

Olivine in Thin section, showing zonal structure crossed polar

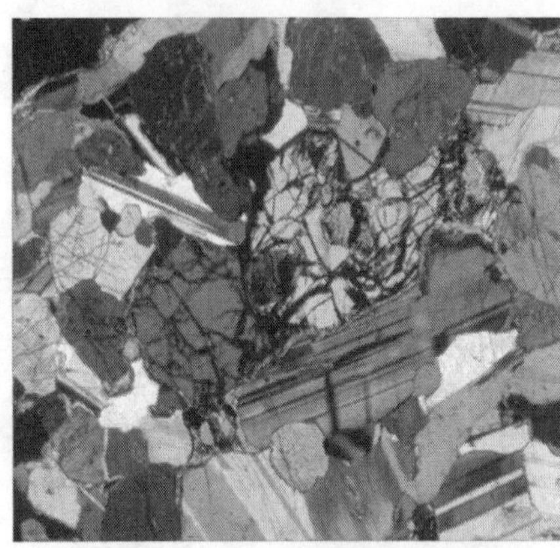

Olivine in Gabbro rock crossed polar section

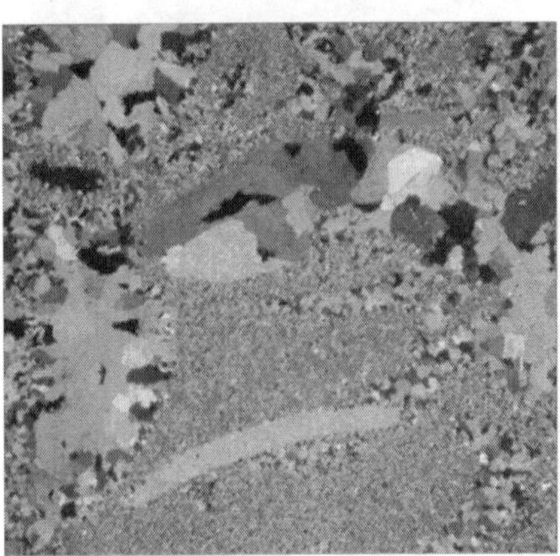

Calcite spar and micrite

Rhombohedral crystals of Dolomite with calcite

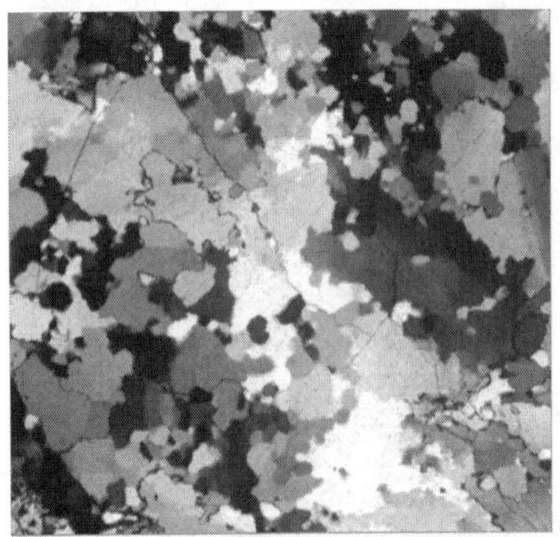

Quartz showing undulose extinction (Wavy)

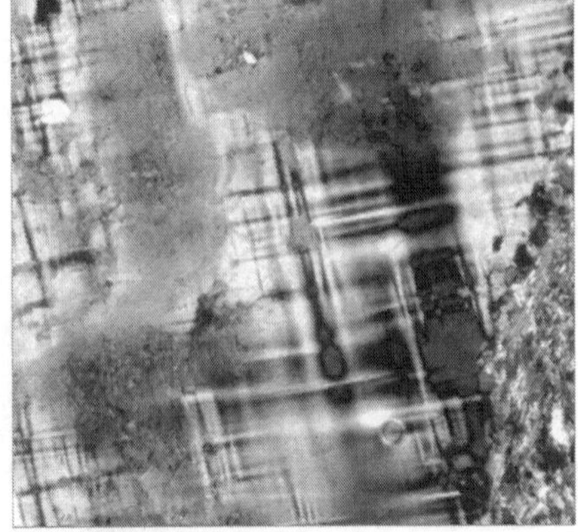

Microcline showing characteristic cross hatching

(*See colour version on page 216*

Plagioclase feldspar in thin section

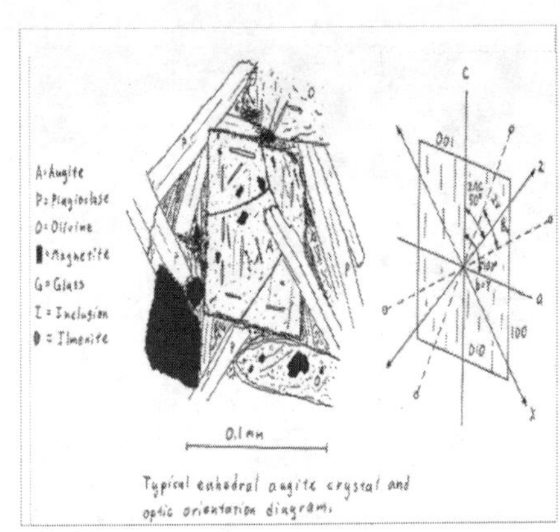

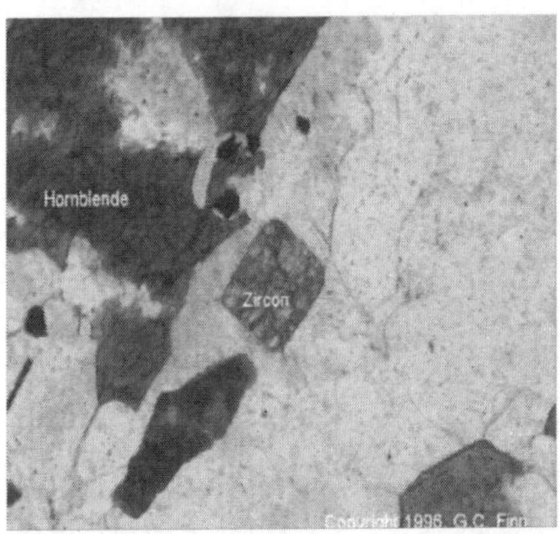

Zircon

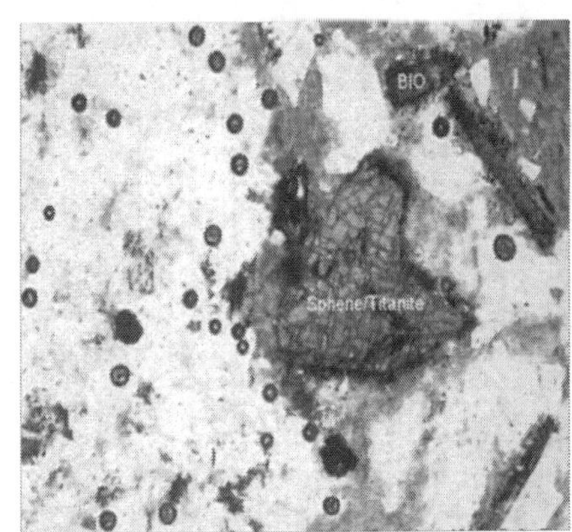

Sphene

Euhedral Garnet in schist

Hornblende
Green hornblende in a diorite. Pleochroism ranges from light yellow-green to bluish-green to brownish-green.

(See colour version on page 217

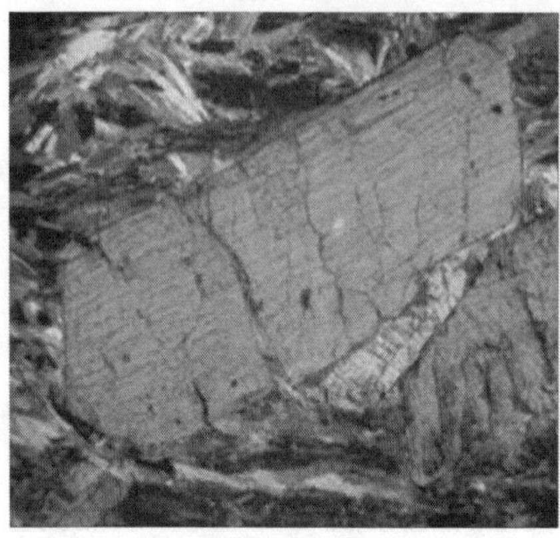

Kyanite in thin section

Thin section, crossed polars: Andalusite (gray) with sillimanite (blue)

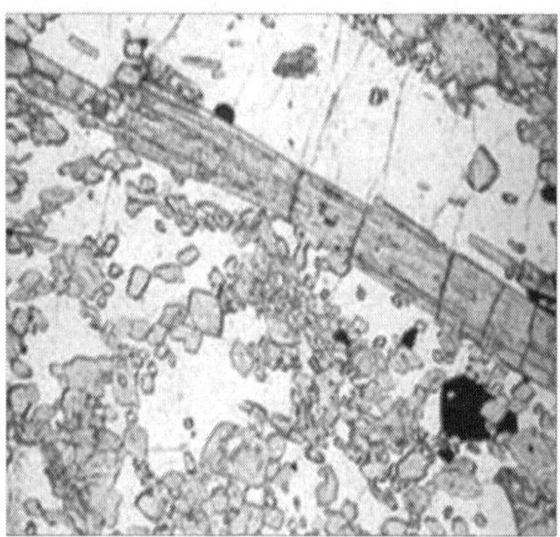

Sillimanite in uncross condition

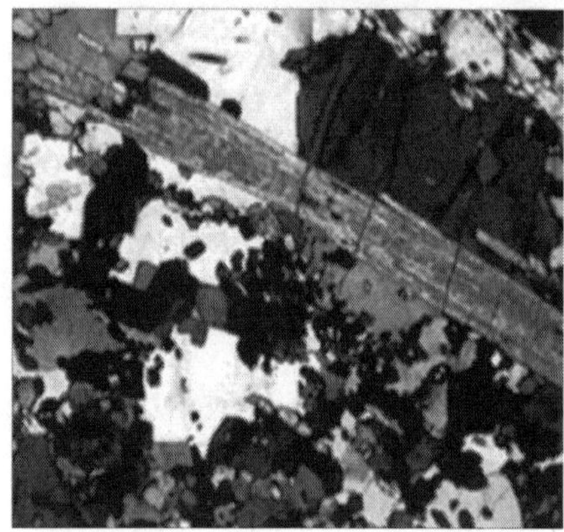

Sillimanite in crossed Nicol condition

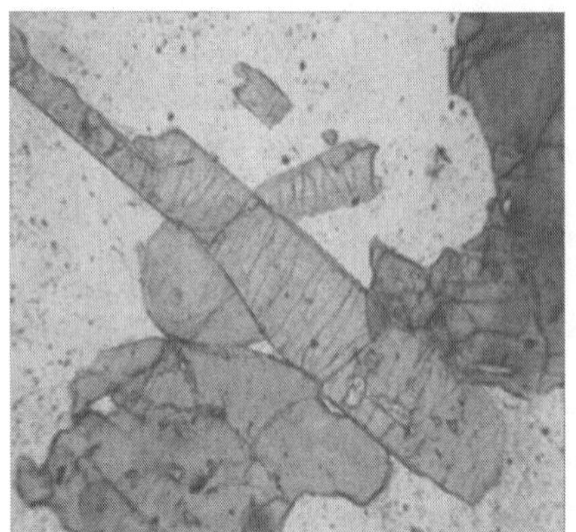

Prismatic crystal of tourmaline in PPL (above) and XPL (Below).

Corundum in thin section

(*See colour version on page 218*

Beryl in plane polarized light

Augite in crossed polar

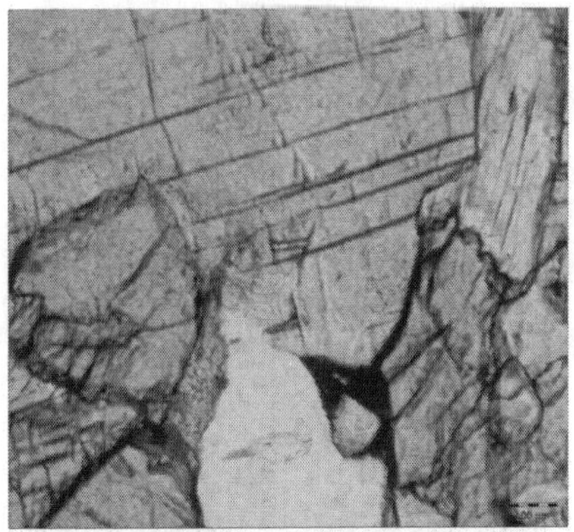

Photomicrograph of diopside in plane polarized light. Diopside is the pale green mineral at the top of the photo; it is surrounded by reddish garnets

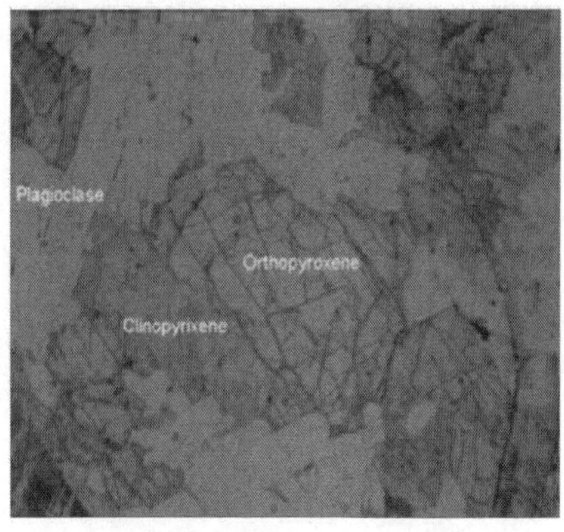

Pyroxene in thin section Ortho and Clino

Biotite
Mineral biotite shows pleochroism in ppl it changes color from dark brown to black when the thin section is rotated

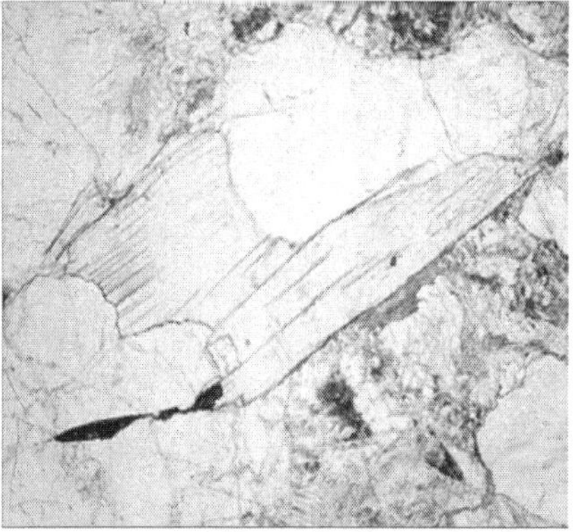

Muscovite,
Muscovite is generally colorless with good cleavage. Pale brown radiation halos can sometimes be visible.

(See colour version on page 219

4.7 Igneous Sedimentary and Metamorphic Rocks

4.7.1 Igneous rocks

Igneous rocks are formed by the consolidation of magma, it is also known as primary rock. Igneous rocks are divided into two types:

a) **Intrusive igneous rock:** These rocks are formed after the consolidation of magma within the layers of earth crust. Example: diorite, gabbro, granite, pegmatite

b) **Extrusive igneous rock:** These ricks are formed after the consolidation of lava on the surface of the earth. Example: andesite, basalt, obsidian, pumice, rhyolite and scoria etc.

4.7.2 Igneous rocks classification

Texture vs. composition

	Felsic	**Intermediate**	**Mafic**	**Ultramafic**
Aphanitic fine grain	Rhyolite	Andesite	Basalt	Conditions needed to produce ultramafic flows do not exist in nature at this time.
Intermediate		Dacite	Diabase	
Phaneritic coarse grain	Granite	Diorite	Gabbro	Peridotite
Glassy	Obsidian			
Frothy	Pumice		Scoria	

4.7.3 Brief description of Igneous rocks

General Rock Type	Texture -ave size of minerals	General color and/or composition-Miscellaneous observations	Rock Name
IGNEOUS ROCKS Interlocking homogenous crystalline texture - no preferred orientation to the mineral grains	Fine grained Extrusive, volcanic	Felsic - Light colored	Rhyolite
		Intermediate	Andesite
		Mafic - Dark colored	Basalt
	Medium grained Dikes, sills, etc.	Felsic - Light colored	------
		Intermediate	Dacite
		Mafic -Dark colored	Diabase
	Coarse grained Generally intrusive	Felsic -Light colored	Granite
		Intermediate	Diorite
		Mafic - Dark colored	Gabbro
		Ultramafic	Peridotite
	Glassy	Dark to black - felsic (DOES NOT follow normal color index)	Obsidian
	Frothy	Felsic -Light colored	Pumice
		Mafic -Dark colored	Scoria

General Rock Type	Texture -ave size of minerals	General color and/or composition-Miscellaneous observations	Rock Name
SEDIMENTARY ROCKS Consolidated detrital clasts, chemical precipitates, and/or biological residue	Coarse Fragments	Rounded clasts	Conglomerate
		Angular clasts	Breccia
	Sand sized fragments	Clean quartz (w/feldspar?)	Sandstone
		Dirty w/rock fragments & clay	Graywacke
	Fine grained - cannot see individual clasts	Nonfoliated, "clay the size"	Siltstone
		Foliated, "clay the mineral"	Shale
	Chemical -fine grain	Soft passes fizz test	Limestone
		Hard - fails fizz test	Chert
	Fossiliferous	Mostly shell fragments	Coquina

General Rock Type		Texture -ave size of minerals	General color and/or composition-Miscellaneous observations	Rock Name
METAMORPHIC ROCKS Interlocking non homogenous crystalline texture commonly with a preferred orientation to the mineral grains	**FOLIATED**	Very fine grained - no visible minerals	Dull - passes "tink test"	Slate
			Foliated, shiny due to increased size of micaceous minerals (almost see them)	Phyllite
		Medium to coarse grain	Individual mineral grains visible. Major mineral(s) included as name modifiers	Schist (ex. Mica Schist)
		Color banded	Alternating layers of light (felsic) and dark (mafic) minerals	Gneiss
		Distinct layering often highly folded and contorted	Alternating layers of felsic igneous rock (light) and mafic gneiss (dark)	Migmatite
		Non-foliated, with non - oriented grains	Soft - passes fizz test	Marble
			Hard - fails fizz test	Quartzite
			Interlocking hornblende crystals	Amphibolite

Source : GeoMan's home page

4.7.4 Brief description of Sedimentary rocks

	Particle Size	Composition	Comments	Rock Name	
Clastic	Coarse Grains > 2 mm	Any rock type (quartz, chert, or quartzite most common)	Rounded clasts	Conglomerate	
			Angular clasts	Breccia	
	Fine Grains 1/16 to 2 mm Can be seen w/ naked eye	Quartz with minor accessory minerals	White, tan, brown; sandpapery feel	Quartz Sandstone	SANDSTONE
		Quartz, with at least 25% potassium feldspar (orthoclase)	Often reddish because of potassium feldspar; sandpapery feel	Arkosic Sandstone	
		Rock fragments, mica, clay, quartz	"Dirty" looking, dark colored; sandpapery feel	Graywacke	
	Very Fine Grains <1/16 mm Cannot be seen w/ naked eye	Quartz and clay minerals	Gritty, 1/16-1/256 mm; some grains may be seen with hand lens	Siltstone	
		Clay sized particles; < 1/256 mm	Non -foliated; "clay the size"	Claystone	
			Foliated; "clay the mineral"	Shale	

Biological	Varies, generally very fine grained	Calcite ($CaCO_3$) fizzes with dilute HCI		Abundant fossils	Fossiliferous Limestone
				Powdery; shells of microscopic plants and/or animals	Chalk
	Coarse to very fine grained	Calcite ($CaCO_3$) fizzes rapidly with dilute HCI		Fossils & fossil fragments; loosely cemented	Coquina
	Varies, generally fine grained	Carbonaceous Material	Plant remains	Brown, soft, porous	Peat
				Brown	Lignite
				Black, sooty, blocky	Bituminous Coal
Bio - Chemical	Generally very fine grained	Calcite ($CaCO_3$)		Non -friable; fizzes with dilute HCI	Limestone
		Dolomite (Ca,Mg) $(CO_3)_2$		Non -friable; fizzes with dilute HCI only when powdered	Dolostone
		Chalcedony (SiO_2)		Light colored, h = 7	Chert
				Dark colored, h = 7	Flint
Chemical	Generally very fine grained	Calcite ($CaCO_3$)	Fizzes with dilute HCl	Medium to coarse grained	Crystalline Limestone
				Ooids (tiny spheres)	Oolitic Limestone
				Banded	Travertine
		Halite (NaCI)		Evaporite, tastes salty	Rock Salt
		Gypsum ($CaSO_4 2H_2 0$)		Evaporite, hardness of 2	Gypsum

4.7.5 Brief description of Metamorphic rocks

	TEXTURE	PARTICLE SIZE	COMPOSITION	COMMENTS	ROCK NAME
Foliated	Foliated	Fine grained, minerals not visible	Minerals with basal cleavage: commonly mica, graphite, etc.	Dense; thin pieces will pass the "tink" test	Slate
			Minerals with basal cleavage: commonly mica, graphite, etc.	Satiny luster; very shiny and reflective in sunlight	Phyllite
	Foliated or Lineated	Medium to coarse grained; grain size generally increases with metamorphic grade, and water is lost	Muscovite, biotite, chlorite, talc, garnet, kyanite, staurolite, feldspar, quartz, tourmaline, and many others	Name is preceded by diagnostic minerals: such as quartz schist, mica schist, quartz mica schist, kyanite biotite hornblende schist, etc.	Schist
	Color Banded		Feldspar, quartz, mica, ferromagnesian minerals	Color banding due to (alternation of light felsic) and dark (mafic) minerals	Gneiss
	Mixed metamorphic and igneous rock		Feldspar, quartz, mica, ferromagnesian minerals	Alternating layers of felsic igneous rock and mafic gneiss; the	Migmatite

Non-foliated	Non-oriented grains	Medium to coarse grained, minerals visible	Calcite ($CaCO_3$)	Hardness of 3; fizzes with dilute HCI	Marble
			Dolomite $(Ca,Mg)(CO_3)_2$	Fizzes with dilute HCI only when powdered	Dolomitic Marble
			Quartz (SiO_2)	H = 7; breaks through grains (as opposed to sandstone that breaks around grains)	Quartzite
			Amphibole; commonly hornblende	Generally black; prismatic crystals with 2 directions of cleavage at 60° / 120°	Amphibolite
			Breaks through clasts as well as around them; clasts may be flattened or stretched	Anything that could be a conglomerate	Meta - conglomerate
		Fine grained, minerals not visible	Clay minerals, mica	Dense, dark colored	Hornfels
			Carbonaceous material	Black, shiny, conchoidal	Anthracite Coal

4.8 Description of Important Intrusive and Extrusive Rocks

4.8.1 Granite

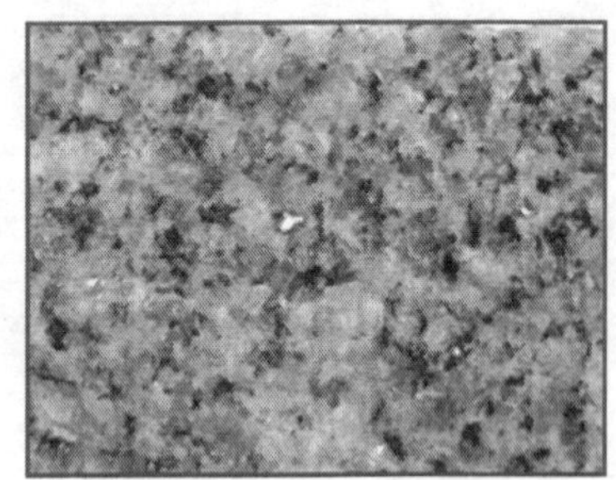

Granite is an intrusive igneous rock with light colour and coarse grained. Texturally, it is holocrystalline that contains interlocking grains of visible quartz, feldspar and mica. Coarse grains are the result of slow cooling of magma in plutonic conditions. Granite occurs as batholithic mass and emplaced as a result of mountain building processes.

4.8.2 Rhyolite

Rhyolite is a volcanic equivalent of granites. It is an extrusive igneous rock with fine grains and light colour. The essential minerals are mainly quartz and feldspar. It is composed of glassyand cryptocrystalline groundmass which sometimes exhibit flow banding.

4.8.3 Obsidian

Obsidian is a volcanic glass resulting from the solidification of molten rocks which cools rapidly so that crystals are not formed. It is extremely glassy modifications of rhyolite. Obsidian is identified by its black colour and conchoidal fracture. Chemically it ranges from oversaturated to saturated rock.

4.8.4 Pumice

(See colour version on page 220)

Pumice is a light colored vesicular rock that is formed by the rapid cooling of a melt where vesicles are formed by the gas trapped in submarine eruption. Gas charged lava cools rapidly and form very porous light weight rock. After submarine eruption, pieces of pumice float on the surface of water and may be carried to great distances by ocean waves.

4.8.5 Scoria

Scoria is a dark colour vesicular extrusive igneous rock, the vesicles are result of emanation of trapped gases. It is basic in composition.

4.8.6 Diorite

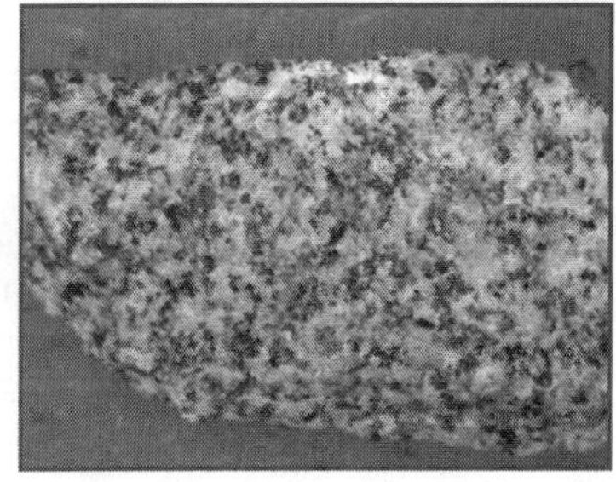

Diorite is a coarse grained intrusive igneous rock. It is intermediate rock with essential mineral, Plagioclase, hornblende. Alkali feldspar and quartz are very less. Texture is holocrystalline or it is said even textured rock. It occurs in the marginal facies of granite. When both plagioclase and alkali feldspar occur in equal proportion, rock is called Adamellite, and when plagioclase feldspar exceeds alkali, it is called Tonalite.

4.8.7 Gabbro

Gabbro is an intrusive, coarse grained, dark coloured basic igneous rock. Mineralogically it is made up of interlocking crystals of feldspar, calc plagioclase, labradorite and some mafic minerals like hypersthenes, augite and sometimes olivine. The texture is course grained holocrystalline intergranular (Gabroic texture). Some gabbro shows ophitic texture in which relatively bigger clino-pyroxene crystals "partially" enclosing prismatic to sub prismatic plagioclase crystals (laths). Gabbro occurs as igneous plutons dyke and sills.

(See colour version on page 220)

4.8.8 Pegmatite

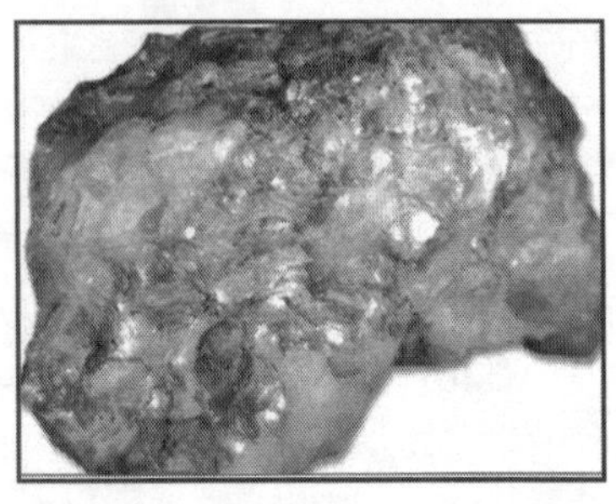

(See colour version on page 220)

Pegmatite is a light colored, extremely coarse grained intrusive igneous rock mostly composed of felspar, quartz, sometimes mica and other rare minerals that is formed near margins of the magma chamber. Pegmatite occurs at hypabyssal condition. It is formed at the final stages from the late magmatic fraction which is enriched in volatiles fluids. The extreme coarseness of mineralogical constituents is due to these volatiles which facilitate movement of nucleating elements.

4.8.9 Andesite

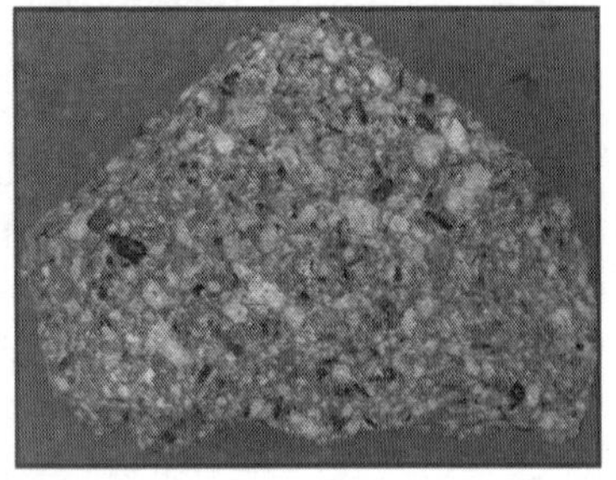

Andesite is an extrusive igneous rock that is fine grained and composed mainly of plagioclase with hornblende, pyroxene and biotite are rarely present. Alkali feldspar is present in subordinate amount, and quartz is negligible. Andesites are formed by consolidation of oversaturated lava and are part of calc alkaline association typical of subduction zone (island arc setup)

4.8.10 Basalt

(See colour version on page 221)

Basalt is a dark colour extrusive igneous rock consists of plagioclase and pyroxene. The term basalt is a Ethiopian word meaning black. Basalt is formed by consolidation of mafic lava. Mafic minerals are augite, clino pyroxene, iron oxide mainly magnetite and ilmenite occur as opaques. Basic magma is much more fluid than acidic lava (rhyolite) so it spreads over vast area. Basaltic magma is saturated to over-saturated in silica, it is called tholeite in which quartz occurs in fine crystalline form. Basalt occurs in large areal extent as flood basalt forming large igneous province e.g. Deccan Trap/

Volcanics (India), Columbia river flood basalt (USA) and Siberia Plateau basalt (Russia).

4.9 Sedimentary Rocks

(See colour version on page 221)

Sedimentary rocks are formed by the deposition and lithification of sediments. These are divided into three basic types.

i) **Clastic sedimentary rocks** are those derived from clastic material which are characteristic of mechanical weathering such as breccia conglomerate sandstone and shale.

ii) **Nonclastic sedimentary rocks** These are derived by precipitation from solution such as salt beds and limestone.

iii) **Organic sedimentary rocks** These are derived from degradation of organic materials (plants and animals) such as coal and some limestone composed of animal shells.

4.9.1 Clastic rocks

i) Rudaceous Rocks

(See colour version on page 221)

When sedimentary rocks are composed of gravel and pebble, cobble or boulder size of clasts. Conglomerates, contain well rounded large fragments/ clasts of rocks. If the pebbles or cobble are mainly of one rock or mineral (generally quartz) it is known as "oligomictic conglomerate". When the pebbles/cobbles are of different rocks it is known as "polymictic conglomerate". Breccias contain angular fragments/ clasts in a finner matrix and indicate low transport. Breccia is heterogenous in composition and may form due to sub-areal weathering. Breccias are classified based upon their origin:

a) **Volcanic Breccia**: When it is formed by angular fragments of volcanic debris, bomb, cemented by volcanic ash.

b) **Fault Breccia/ Crush Breccia**: When it is formed by movement of rocks along fault plane. it is the initial stage of mylonitisation.

c) **Collapse Breccia**: When the roof of a cave collapses and, the angular fragments with the matrix are cemented, then it forms Collapse Breccia.

ii) **Shale**: It is derived from finer sediments (clay size 1/256 millimetre plains). It is fine laminated sedimentary rock, splits easily along bedding planes. It is difficult to identify mineral in hand spacimen. Shale is mainly composed of hydrated silicates of aluminium, or hydrated oxides of iron. Sometimes it contains calcium carbonate, Iron sulphates. The argillaceous rocks are products of weathering, and deposition under quiet condition, in the deeper parts of ocean. Shale exhibits fine laminated, close spaced bedding, but in **Mudstone**, the bedding is less developed because it is mainly composed of un-consolidated clayey material.

iii) **Sandstone** is made up of sand size particles, 1/16 to 2 millimetre, which are deposited in environments where such sand can accumulates.

(See colour version on page 221)

iv) Siltstone: It is a clastic sedimentary rock which is made up of sediments size between 1/256 and 1/16 millimetre.

(*See colour version on page 221*)

4.9.2 Non clastic rocks

Rocks formed by chemical or biochemical processes are known as non-clastic rocks. They are divided into two groups:

i) Evaporitic

ii) Non evaporitic

i) Evaporitic

Salt : It is a sedimentary rock which is derived by the evaporation of ocean water or saline lake. It is also called as halite mineral.

ii) Non Evaporitic

Limestone: It is a sedimentary rock which is produced by both precipitation of chemically dissolved calcium carbonate and from the accumulation of organic shells.

Dolomite: It is a sedimentary rock which is also known as dolostone or dolomite rock. Dolomite is formed in similar way as limestone but contains magnesium which is derived from the sediments by the circulating waters.

Coal: It is a organo-chemical sedimentary rock that is made up of plant remains deposited in shallow basins on the continents.

(*See colour version on page 222*)

4.10 Metamorphic Rocks

Metamorphic rocks are those rocks which have been modified by heat, pressure and chemical process in the subsurface. These classified as 1) Foliated metamorphic rocks such as slate, phyllite, schist and gneiss which show distinct layering or banding. 2) Non-foliated metamorphic rocks such as marble, hornfels and quartzite and do not have layers or bands.

i) **Gneiss**: It is a foliated metamorphic rock that has bands of light and dark colours. The lighter mineral bands show intergranular texture and contains abundant quartz and feldspar minerals (light) with dark (mica and other platy minerals) Example: Granite Gneiss.

ii) **Phyllite** : It is a foliated metamorphic rock that is made up of fine grained mica (mainly sericite). It is intermediate between slate and schist.

iii) **Schist** : It is a metamorphic rock with well developed foliation and contains significant amount of mica which allows rock to split along the foliation into thin sheets.

iv) **Slate** : It is a low grade metamorphic argillaceous rock which splits along thin partings into platy rectangular pieces.

(See colour version on page 222)

v) **Marble** : It is a non-foliated metamorphic rock which is crystalline and composed mainly of calcite. It is derived from thermal metamorphism of limestone/calcarious sediments.

(See colour version on page 222)

vi) **Amphibolite** : It is a non-foliated metamorphic rock formed by the recrystallization of Basic rock or sometimes argillaceous under conditions of high temperature and directed pressure. In case of its derivations from basic igneous rocks, it is mainly composed of Amphibole or plagioclase with minor quartz which shows distinct alignment of grains (foliation). In case of derivations from calc-magnecian pelitic rock, the amphebolite may contain apart from hornblende, actinolite also depending upon the composition of original rock and grades of metamorphism.

(See colour version on page 223)

vii) **Hornfels :** It is a fine grained non-foliated metamorphic rock produced by contact metamorphism. Hornfels varies in composition depending upon the rock affected by contact metamorphism. This shows typical 'HORN' like sheen.

viii) **Quartzite** : It is a non-foliated metamorphic rock produced by metamorphism of sandstone which is mainly composed of quartz grains (Siliciclastic).

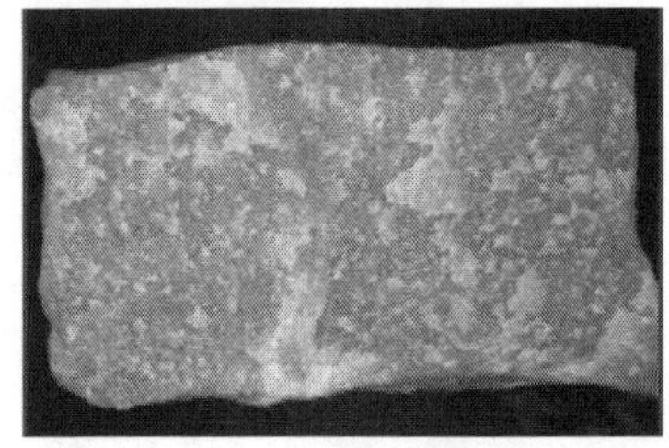

(See colour version on page 223)

4.11 Microscopic Identification of Different Rocks

Igneous Rock: Granite

Granite is a acidic igneous rock. It contains mainly quartz and alkali feldspar as essential minerals. Microscopically it shows holocrystalline texture, formed by the slow cooling of magma in plutonic condition. In the photomicrograph at right of centre it shows an intergrowth texture, it has quartz and feldspar forming an intergrowth - pairs of crystals grown through each other in a complex pattern. (Field of view 8 mm, polarizing filters). Wavy extinction of quartz is very clear.

Granite in thin section

(See colour version on page 223)

Gabbro (olivine-gabbro)

Gabbro is a basic igneous rock, mainly made up of dark and light colored minerals. Half the rock is made of minerals (pyroxene and olivine) that are dark-colored in hand specimen. In crossed nicol condition, they show bright colours. The striped grey rectangular crystals are plagioclase feldspar. (Field of view 8 mm, polarising filters).

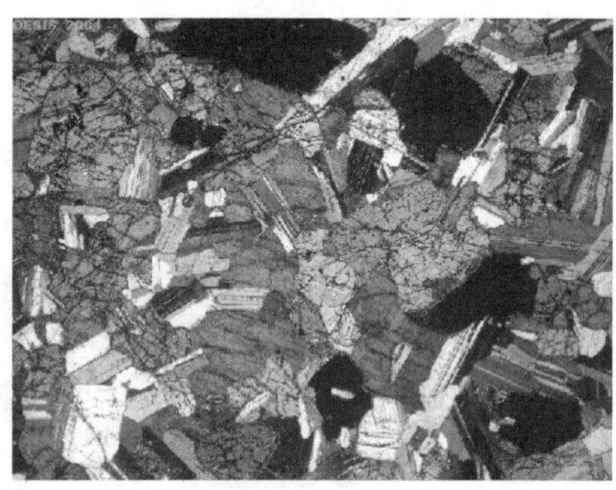

Gabbro in thin section

(See colour version on page 223)

Basalt (olivine basalt)

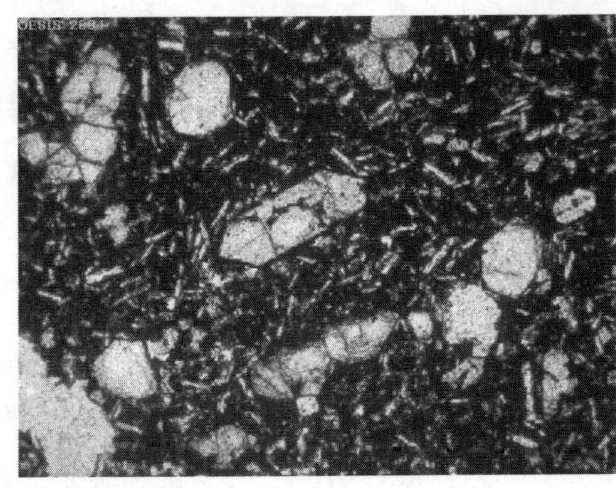

Basalt in thin section

(See colour version on page 223)

Basalt is a basic volcanic rock In this thin section, the rock is manly composed of olivine and plagioclase feldspar. It has larger crystals in a very fine-grained matrix (porphyritic texture). The large crystals are phenocryst of olivine grew while the magma was held in a chamber beneath the volcano, but the rest of the rock crystallized only after eruption when the lava flow quickly cooled. Fine groundmass is madeup of plagioclase seen as laths or needle shape, and pyroxene. (Field of view 3 mm. under crossed nicols)

Sedimentary rock

Sandstone

Sandstone in thin section

(See colour version on page 223)

Sandstone is a sedimentary rock made of quite well rounded grains of quartz, cemented together by any of the cementing material like silicious, arenaceous/calcarious/ferruginous/carbonaceous /argillaceous, the roundness of quartz grains represent it's movement from the source area. Field of view 3.5 mm, polarizing filters.

Limestone (dolomitic)

Dolomitic limestone in thin section
(*See colour version on page 223)*

Limestone is a sedimentary rock, formed by the chemical or organo-chemical process of deposition. In this limestone, rhomb-shaped crystals of dolomite (calcium magnesium carbonate) have grown after deposition, while the sediment was being changed into rock. They replace the fine calcium carbonate mud (dark material in the photo) that makes up the rest of the rock. Dolomitization process in limestone enhances it's porosity and then the rock is considered as good aquifer for petroleum. Field of view 3.5 mm.

Slate

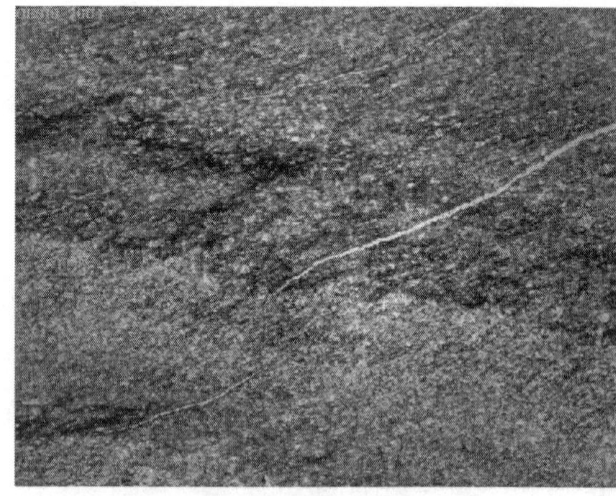
Slate in thin section
(*See colour version on page 224)*

Slate is a metamorphic rock, and is formed from fine-grained sediments such as mudstone. When these are compressed and heated a little, tiny new flakes of mica mainly sericite grow, and tend to align themselves at right angles to the direction of compression. Although the individual mica crystals cannot be seen, the rock breaks along a particular direction, or cleavage plane. Here you can see the cleavage, and can also observe that it is not parallel to the original bedding marked by dark and light bands under crossed nicols. Field of view 2.5 mm.

Phyllite

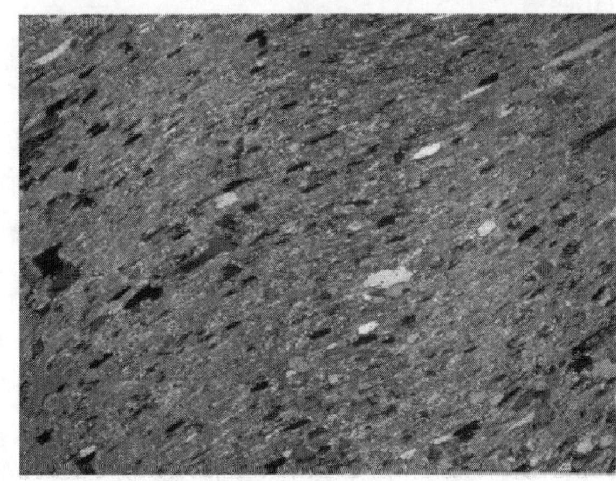
Phyllite in thin section

(See colour version on page 223)

Phyllite is a metamorphic rock, similar to a slate, except that it forms at higher temperatures and pressure. Now the new mica flakes are large enough to be seen under the microscope, and form mats of crystals (pink when seen between crossed polarisers) orientedparallel to each other. In hand specimen this rock shows glossy sheen, but individual mica flakes cannot be distinguished with the naked eye. Field of view 2.5 mm, (under crossed nicols).

Schist (mica schist)

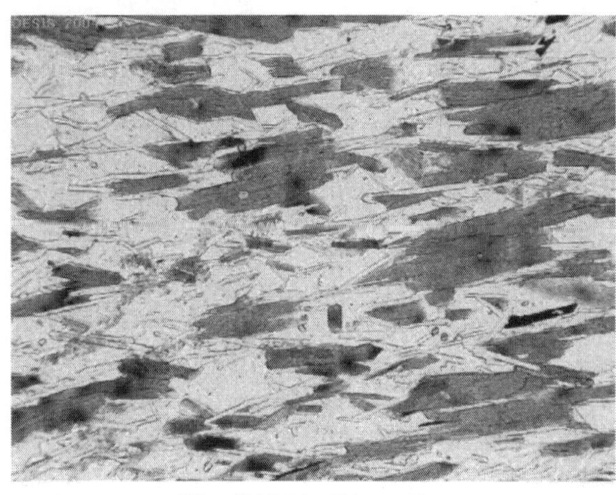
Mica Schist in thin section

(See colour version on page 224)

Schist is a high grade metamorphic rock, formed by regional metamorphism of argillaceous sediments.. It contains platy tabular and rod like minerals like mica, chlorite, talc, amphibole etc. Due to metamorphic effect all the platy, tabular minerals arranged parallel to each other, perpendicular to compressional stress a structure known as schistosity-means the rock splits along these planes. In this thin section, both brown and colorless mica flakes can be seen aligned. Field of view 1.5 mm (under crossed nicols).

Gneiss

Gneisses are highly metamorphosed rocks that have banding or alignment of alternate platy and granular minerals which, is known as gneissose structure. In this thin section, plagioclase, hornblende, quartz and biotite are arranged in alternate linear manner as platy and granular minerals. Twinning in plagioclase and wavy extinction in quartz is very clear (field of view is under crossed nicols).

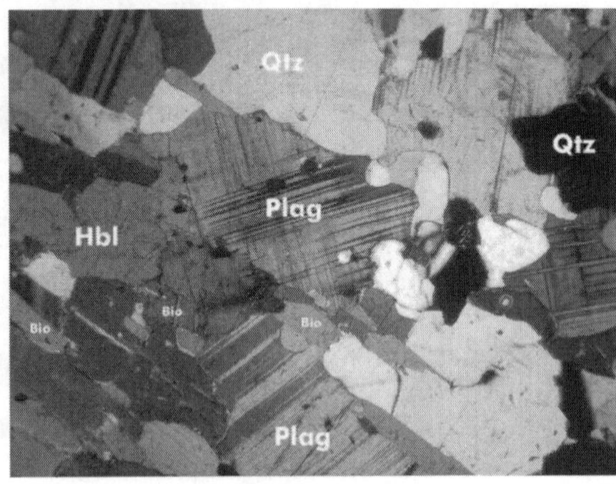

Gneiss in thin section

(*See colour version on page 224)*

Marble

Marble is a metamorphic rock, formed by the thermal metamorphism of limestone. The calcium carbonate reforms itself into larger, interlocking crystals of calcite (e.g. the pearly-colored crystals in the centre). The impurities are converted into new metamorphic minerals. In this case, the larger bold-colored crystals are forsterite (magnesium silicate, a variety of olivine). Field of view 6 mm, (under crossed nicols).

Marble in thin section

(*See colour version on page 224)*

4.12 Megascopic and Microscopic description of Rocks

4.12.1 Igneous rocks

1. Colour - Leucocratic / Mesocratic / Melanocratic.
2. Texture: (for megascopic and microscopic both)
   - Crystallinity - Holocrystalline/Merocrystalline/Holohyaline
   - Granularity
     - Phaneric - Coarse / Medium / Fine grained
     - Aphaneric - Microcrystalline / Cryptocrystalline
   - Shape of Grain - Euhedral / Subhedral / Anhedral
   - Mutual relationship
     - Equigranular - Panidiomorphic/Hypediomorphic/Allotrimorphic
     - Inequigranular - Porphyritic/Poikilitic/Ophitic/Intergrowth. Intergranular/Intersertal/Directional-Trachytic
3. Mineral composition
4. Mode of Occurrence/Origin - Plutonic/Hypabyssal/Volcanic
5. Name of Rock - ....
6. Indian Localities - ....
7. Economic Use - Building stone / Road material / Host rock of ore minerals

4.12.2 Sedimentary rocks

1. Color - Light / Medium / Dark colored rock
2. Texture
   - Grain size - Boulder/Cobble/Pebble/Sand/Silt/Clay.
   - Grain sorting - Well sorted / ILL sorted
   - Grain shape - Rounded/Angular

3. Cementing material
   - Silicious/ Arenaceous - If made up of silica cement
   - Calcareous - If made up of calcite cement
   - Ferruginous - If made up of ferruginous material
   - Carbonaceous - If made up of carbon material
   - Argillaceous - If made up of clay size material
4. Mineral composition - Ex. Quartz for sandstone/ Qtz +Felspar for Arkose
5. Structure if any - Bedding/ ripple marks/ Current bedding/ Rainprint/Fluite mark/Mud cracks/ conglomerate/stromatolitic/oolitic/
6. Origin/Mode of Occurrence - Mechanical/Chemical/Biogenic
7. Indian Locality - ....
8. Economic Use - Building stone/Road material/Host rock of ore minerals

4.12.3 Metamorphic rocks

1. Color - Light/Medium/Dark colored rock
2. Texture
   - Crystalloblastic/Palimpsest/Porphyroblastic/Blastoporphyritic etc.
   - Structure: Cataclastic/Maculose/Granulose/Schistose/Gneissose
3. Foliated/Non foliated
4. Mineral composition
5. Origin/Mode of Occurrence - Mechanical/ Chemical/ Biogenic
6. Indian Locality - ....
7. Economic Use - Building stone/Road material/Host rock of ore minerals.

❑❑❑

Chapter **5**

Palaeontology

The remains of animals and plants of past ages preserved in the rocks are known as fossils, the study of which forms the subject of Palaeontology. The conditions for an animal or plant to become a fossil are firstly it must possess skeleton of some kind or some hard parts, since the soft parts are decomposed rapidly, secondly the organism must be covered up by sedimentary layer, otherwise it will soon crumble to pieces.

5.1 Modes of Preservation of Fossils

The conditions in which fossils occur depends, on their original composition and on the material in which they are embedded. The chief types are as follow:

5.1.1 The entire organism preserved

Occasionally the soft parts of the organism are preserved as well as the skeleton, the whole having suffered very little change. Instances of this are the woolly rhinoceros and mammoth found frozen in the mud and ice in Northern Siberia. Insects encased in fossil resin, known as amber, are found in the Oligocene beds in the Baltic shores of Persia, but the soft part of organisms are rarely preserved.

5.1.2 The skeleton preserved almost unchanged

Sometimes where the skeleton alone is preserved, it remains almost in its original condition, except that it has lost its organic matter. Thus the shells in the Pliocene beds of England differ from living ones only in being lighter, more porous and generally colourless. In some instances a certain amount of mineral matter, such as carbonate of lime, has been added to the skeleton, making it heavier and more compact.

5.1.3 Carbonisation

In some plants and in animals with chitinous skeletons, such as graptolites, the original material usually becomes carbonized when the organism undergoes decomposition and loses oxygen and nitrogen, the relative percentage of carbon therefore increases. The changes are similar to those which occurred during the conversion of vegetable matter into coal.

5.1.4 Mold and Cast

Sometimes the skeleton or hard parts disappears entirely, a mould only remains it is especially the case when it consists of aragonite and is embedded in a porous stratum. The mold resembles like original fossil. The impression of inner side of hard part is known as internal mold while impression of outside of hard part is called external mold. A cast is formed by the filling of mold.

5.1.5 Petrification

In some the fossils show minute structure as well as the form of the organism, but the original material of the skeleton has been replaced by another mineral. Thus we find fossil wood which shows the cells and vessels just as in existing trees, but in which the walls are formed of silica instead of cellulose.

5.1.6 Imprints

The footprints of animals and the impression is of jelly-fishes are sometimes found in the rocks, and these, although forming no part of the animal itself, are nevertheless regarded as fossils (trace fossils).

5.2 Description of Fossils

5.2.1 Phylum Protozoa

Phylum Protozoa contains organisms of lower form, i.e. the body is consists of one cell only. Based on morphology, Protozoa have been divided into four classes; but only two classes 1) Gymnomyxa and 2) Flagillata contain organism in fossil state.

Description of Fossils

Fossil Name	:	Nummulites
Classification		
Phylum	-	Protozoa
Class	-	Saracodina
Order	-	Foraminifera
Name	-	Nummulites

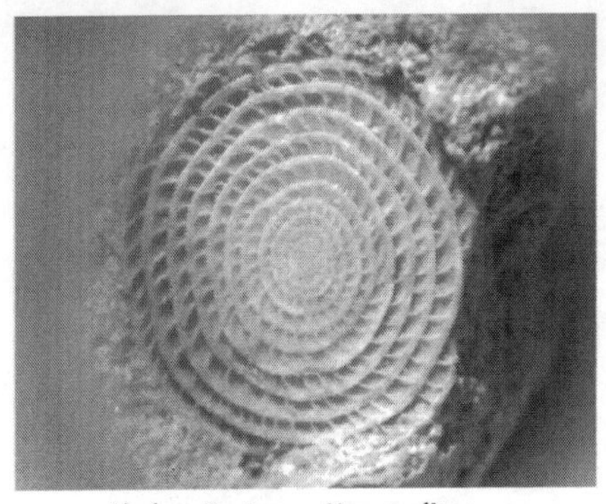

Phylum Protozoa Nummulites

Phylum Coelenterata Calceola

Phylum Coelenterata Zafrentis

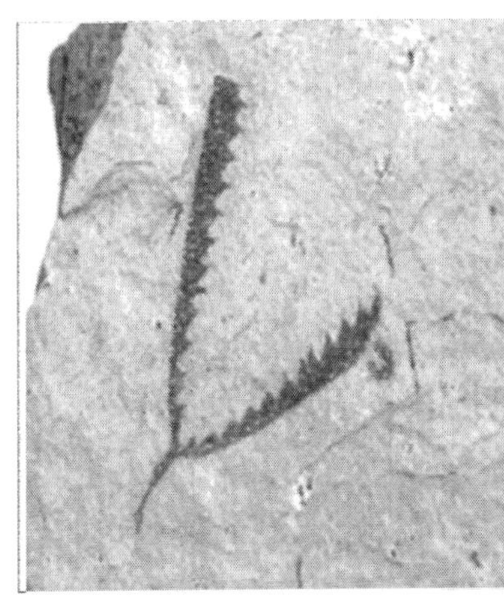

Monograptus (Graptolites)

Diplograptus (Graptolites)

(See colour version on page 225)

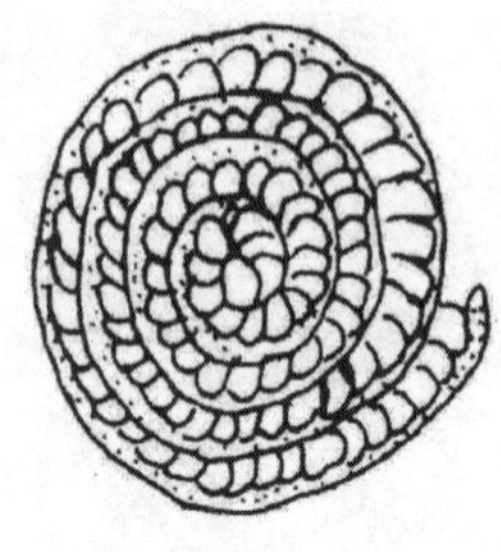

NUMMULITES
(Order : FORAMINIFERA)

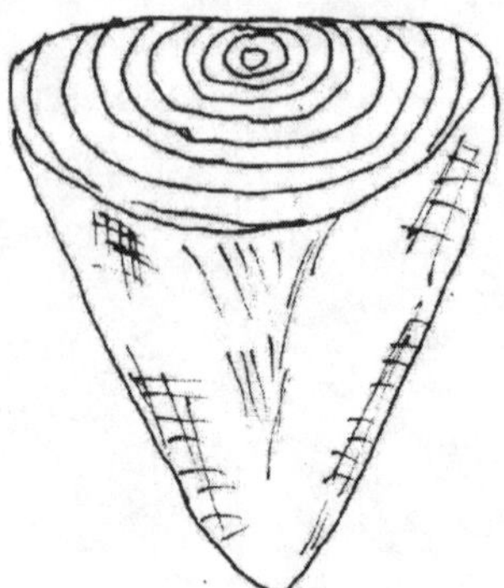

CALCEOLA
(Order : RUGOSA)

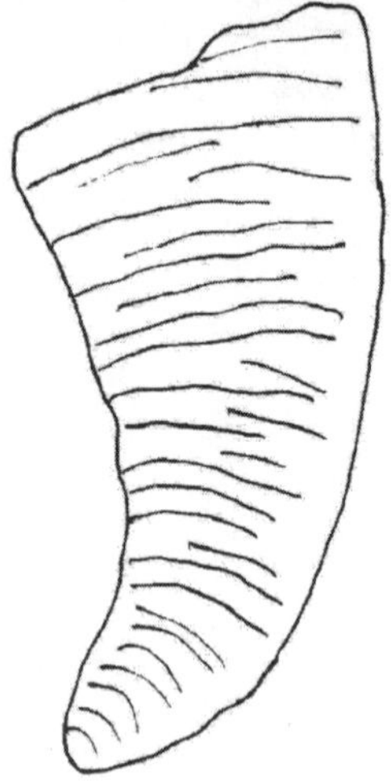

ZAPHRENTIS
(Order : RUGOSA)

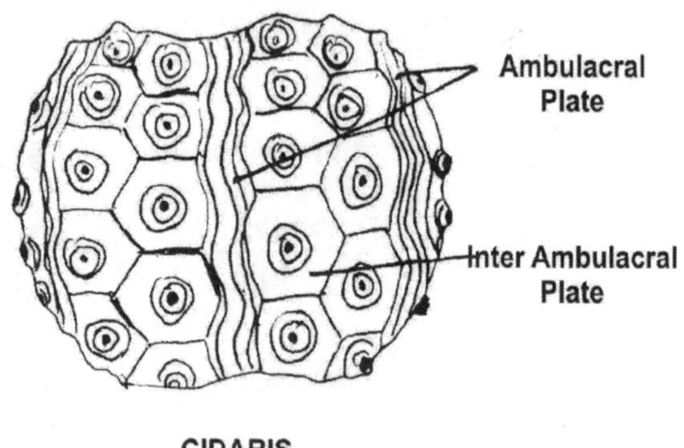

CIDARIS
(Ambulacral Plates)

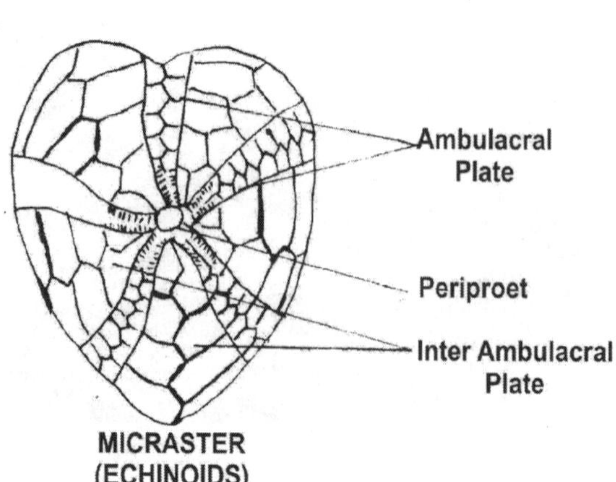

MICRASTER
(ECHINOIDS)

A. Morphology

1. Test lenticular, composed of circular rings that are divided by septa
2. The inner circle is completely covered by outer circles
3. These fossils are dimorphic (two form of same species)
4. One having larger spherical centre called Megalosphere and other having much smaller centre called microsphare
5. Whorls are divided into chambers by septa
6. Each septum is separated by imperforate laminal.

B. Geological Age

Eocene to Oligocene, Nummulites are used as an indicator of petroleum and gas occurrence.

5.2.2 Phylum Coelenterata

Phylum Coelenterata includes corals, sea anemones, jelly fishes etc. Morphologically they exhibit radial symmetry with one internal cavity. In the living stage, the body wall consists of an outer layer ectoderm and an inner layer endoderm. Phylum Coelenterata is divided into four classes: i) Hydrozoa; ii) Seyphozoa; 3) Anthozoa; 4) Ctenophora. Organisms from class Hydrozoa and Anthozoa are only contains fossil.

Description of fossils

Fossil Name	-	Graptolites
Classification		
Phylum	-	Coelenterata
Class	-	Hydrozoa
Order	-	Graptolithina
Sub Order	-	Graptolitoidea
Name	-	Graptolites (Monograputs, diplograptus, Tetragraputs.

Morphology Graptolite

1. The entire skeleton of graptolite is termed as rhabdosome.
2. It is unbranched in Monograptus and branched in Diplograptus/ Tetragraptus.
3. The row of small cells are called hydrotheca. Each cell opens in common canal.

Geological Age: Lower Palaeozoic.

II. Description of fossil

Fossil Name Calceola

Classification

Phylum Coelenterata

Class Anthozoa

Order Zoantharia

Sub Order Rugosa

Name Calceola

Morphology:

1. Shell is conical, one side is flat other is convex.
2. Calyx is very deep and closed by operculum.
3. Septa indicated by streae in the calyx, wall thick.

Geological Age: Middle Devonian.

III. Description of fossil

Fossil Name Zaphrentis

Phylum Coelenterata

Class Anthozoa

Order Zoantharia

Sub Order Rugosa

Name Zaphrentis

Morphology:

1. Shell conical curved, Calyx deep.
2. Theca thick, Septa numerous large reaching to centre.
3. Tabulae well developed. Columella absent.

Geological Age: Devonian and Carboniferous.

5.2.3 Phylum Brachiopoda

The shell of brachiopod consists of two valves (inequivalve), different in sizes called the dorsal and ventral, the valves show bilateral symmetry. Based on morphology, Phylum Brachiopoda have be divided in two classes: Inarticulata and Articulata. In Inarticulata, the two valves are held together by muscles, while in Articulata, the central valve has a pair of hinge teeth that fit into slots on the underside opposite the hinge line of the dorsal valve. Most brachiopods attach to the substrate by means of a cylindrical extension called pedicle.

Description of fossils

1. Products

Classification

Phylum	-	Brachiopoda
Class	-	Articulata
Order	-	Protremata
Name	-	Productus

A. Morphology

- Shell is free or fixed by spines. Hinge line straight and teeth absent
- Surface ornamented with radiating ribs crossed by concentric folds
- Brachial skeleton not present

B. Geological Age - Carboniferous and Permian (Extinct)

2. Spirifer

Classification

Phylum	-	Brachiopoda
Class	-	Articulata
Order	-	Telotremata
Name	-	Spirifer

A. Morphology

Shell transfers more or less triangular, biconvex and ornamented by radiating ribs. Hinge line is straight and long. There is a sinus in ventral value and a ridge on the dorsal value.

Brachiopod Morphology

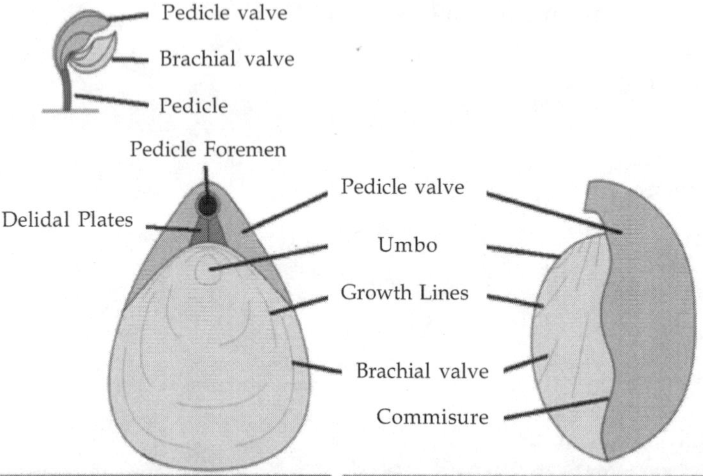

Pedicle valve
Brachial valve
Pedicle

Pedicle Foremen
Delidal Plates
Pedicle valve
Umbo
Growth Lines
Brachial valve
Commisure

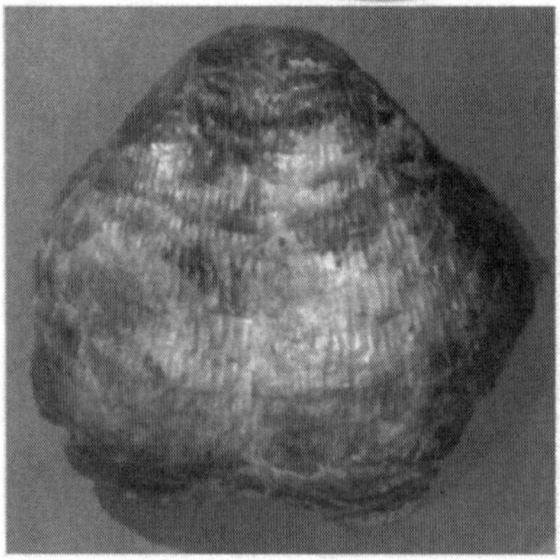

Productus

Spirifer

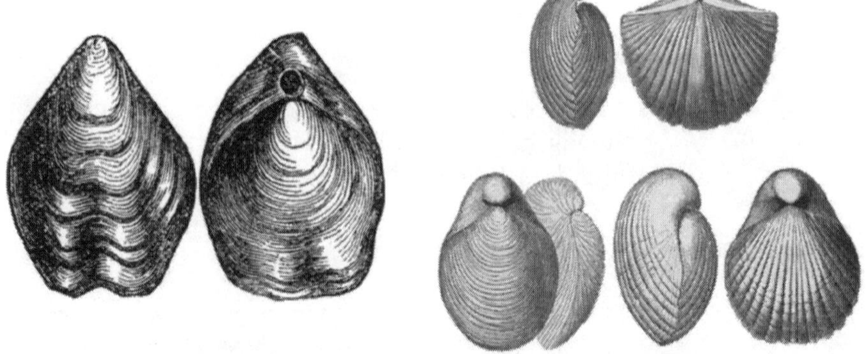

Terebratulla

Rhynconella

(*See colour version on page 226*)

B. Geological Age - Silurian to Permian

3. Rhynconella

Classification

Phylum	-	Brachiopoda
Class	-	Articulata
Order	-	Telotremata
Name	-	Rhynconella

A. Morphology

1. Shell triangular or pyramidal in shape
2. Dorsal value having a corresponding fold surface exhibiting radiating ribs
3. Teeth large, crenulated dental plate short
4. Adductor impression oval. Muscular impression small
5. Crura narrow slightly curved towards the ventral value

B. Geological Age - Devonian to present

4. Terebratulla

Classification

Phylum	-	Brachiopoda
Class	-	Articulata
Order	-	Telotremata
Name	-	Telebratulla

A. Morphology

1. Shell biconvex, oval, elongate rounded
2. Shell consists of two folds on dorsal valve and sinus on the ventral surface is smooth
3. Hinge line is curved, dental plate is absent
4. Brachial skeleton present
5. On the dorsal value strong adductor impression occurs widely separated

B. Geological Age – Eocene to Pliocene

5.2.4 Phylum Mollusca

The majority of the molluscs (oysters, cuttlefish, etc.) are marine, but some live on land, others in fresh water. Unlike the worms and arthropods, they are un-segmented animals, and bear no serially repeated appendages. The body is bilaterally symmetrical and there is a repetition of the same organs on each side. Phyllum Mollusca can be divided into three classes.

Classes: - Lamellibranchia
- Gastropoda
- Cephalopoda

I. Class Lamellibranchia

In the lamellibranchia, unlike brachiopod, the shell is consists of two valves, dorsal and ventral, equal in size (equivalve) and placed one on the right, the other on the left side of the body. Both are joined together by means of hinge and ligament and adductor muscles.

Description of fossils

1. ARCA

Phylum - Mollusca
Class - Lamellibranchia,
Order - Taxodonta

A. Morphology

i) Shell equivalve, surface ornamented with radiating ribs and concentric growth lines: margins smooth or dentate; gaping ventrally

ii Hinge straight, with numerous, small similar, transverse teeth (Taxodonta)

iii) Umbo prominent, separated by the large areas, which have numerous ligamental grooves converging from the hinge-margins to below the umbo

iv) Adductor impressions sub-equal, the anterior rounded, the posterior divided. Pallial line simple

B. Age : Jurassic to present day

2. PECTEN

Phylum	-	Mollusca
Class	-	Lamellibranchia
Order	-	Anisomyaria
Sub-order	-	Pectinacea

Morphology

i) Shell sub-circular, ovate or trigonal, closed, almost equilateral, nearly equivalve

ii) Surface ornamented with radiating ribs or striae, sometimes smooth or with concentric ridges

iii) Hinge-line straight; with well-developed ears, a central, triangular pit for the internal ligament

iv) Adductor impressions large, a little excentric

B. Age: Carboniferous to present day

3. OSTREA

Phylum	-	Mollusca
Class	-	Lamellibranchia
Order	-	Anisomyaria
Sub-order	-	Ostreacea

A. Morphology

i) Shell with lamellar structure, irregular, inequivalve, slightly inequilateral, fixed by the left (larger) valve. Left valve convex, often with radiating ribs or striae.

ii) Umbo prominent, sometimes directed anteriorly, sometimes posteriorly.

iii) Right valve flat or concave, often smooth. Ligament-pit triangular or elongated.

iv) Hinge short, without teeth. Adductor impression sub-central; pallial line indistinct. No foot.

Lamellibranchia Morphology

Venus

Cardita

Exogyra

Arca

Ostrea

Gryphaea

Pecten

(See colour version on page 227)

B. Age. Triassic to Present

4. GRYPHAEA

Phylum	-	Mollusca
Class	-	Lamellibranchia.
Order	-	Anisomyaria
Sub-order	-	Ostreacea

A Morphology

i) Shell irregular, inequivalve, fixed by the left large valve in the young stage only, free in the adult;

ii) Left valve large and convex, with a prominent incurved umbo, while right valve flattened or concave.

iii) Gryphea have originated independently from more than one species of Ostrea.

B. Age : Lias (Early Jurassic)

5. EXOGYRA

Phylum	-	Mollusca
Class	-	Lamellibranchia
Order	-	Anisomyaria
Sub-order	-	Ostreacea

A. Morphology

i) Shell irregular, inequivalve

ii) Fixed by the left (larger) valve. Right valve flat, resembling an operculum.

iii) Umbo is more or less spiral, directed posteriorly

B. Age: Upper Jurassic to Chalk (Late Cretaceous)

6. CARDITA

Phylum	-	Mollusca
Class	-	Lamellibranchia
Order	-	Eulamellibranchia
Sub order	-	Heterodonta

A. Morphology

i) Shell ovoid or oblong, elongated, in-equilateral, with broad radial ribs.

ii) Umbones prominent, anterior. Lunule small. Ligament external.

iii) Right valve have two long, parallel cardinal teeth, nearly horizontal, and a very small anterior lateral tooth. The dentition type is called Heterodont.

iv) In the left valve one short anterior cardinal, and one long posterior cardinal tooth.

v) Adductor impressions large. Pallial line simple. Margins of valves coarsely crenulate.

B. Age: Trias to present day

7. VENUS

Phylum	-	Mollusca
Class	-	Lamellibranchia
Order	-	Eulamellibranchia
Sub-order	-	Heterodonta

A. Morphology

i) Shell oval, convex, ornamented with concentric lamellae, or with radiating ribs;

ii) Lunule distinct. Margins of valves crenulated.

iii) Hinge-plate is wide; each valve contains three thick cardinal teeth, lateral teeth absent.

iv) Ligament is prominent. Pallial sinus is angular and short.

B. Age: Miocene to Present

II. Class Gastropoda

The class Gastropoda consists of mollusks with a univalve (one piece) shell, which is coiled in a helicoids spiral (Snail). Like lamellibranchs the shell is composed of aragonite. Gastropods are mainly marine and banthonic organism, they crawl over the sea floor. Some genera live in fresh water (pond snails) a few also reported as land snails, which live on land and breathe air.

Description of fossils

1. TURBO

Phylum	-	Mollusca
Class	-	Gastropoda
Order	-	Prosobranchea

A. Morphology

i) Shell turbinate solid, conical, whorls convex, interior nacreous.

ii) Aperture large, circular, entire, slightly projected anteriorly ;

iii) Outer lip sharp. Columella curved, flattened. Imperforate / with a small umbilicus.

iv) Operculum thick, calcareous, exterior convex, interior flat and spiral, nucleus central or sub-central.

B. Age: Jurassic to Present

2. NATICA

Phylum	-	Mollusca
Class	-	Gastropoda
Order	-	Prosobranchia
Sub order	-	Pectinibranchea

A. Morphology

i) Shell smooth, globular, spire short, Body whorl bigger than preceding one.

ii) Aperture semi –oval entire; outer lip sharp, oblique; inner lip thickened with callus, not crenulated.

iii) Umbilicus present, partly covered by callus. Operculum of the same size as aperture, horny or calcareous.

B. Age: Trias to Present

3. TURRITELLA

Phylum	-	Mollusca
Class	-	Gastropoda
Order	-	Prosobranchea
Sub order	-	Pectinibranchea

A. Morphology

i) Shell turreted with many flat or slightly convex whorls
ii) Umbilicus absent. shell ornamented with spiral ribs
iii) Spire very long and acute. Aperture oval or sub -quadrate, entire
iv) Outer lip thin, sinuous, slightly produced in front. Operculum horny

B. Age: Cretaceous to Present

4. CERITHIUM

Phylum	-	Mollusca
Class	-	Gastropoda
Order	-	Prosobranchea
Sub order	-	Pectinibranchia

A. Morphology

i) Shell turreted, umbilicus absent
ii) Whorls many, narrow, last whorl much shorter than the spire
iii) Spiral ornamented with numerous slightly grooved lines running parallel to whorl
iv) Aperture semi-oval, with a short posterior canal
v) Outer lip thickened and reflected
vi) Columella edge concave
vii) Operculum thorny, oval, with submarginal nucleus.

B. Age: Cretaceous to Present

5. MUREX

Phylum	-	Mollusca
Class	-	Gastropoda
Order	-	Prosobranchea
Sub order	-	Pectinibranchea

A. Morphology

i) Shell thick, elongate, spire prominent and sharp

ii) Whorls convex, spiny, foliaceous, or tubercular

iii) Aperture ovate anterior canal long, straight tubular, nearly closed, posterior canal absent

iv) Outer lip thick, inner lip smooth. Operculum oval.

B. Age: Eocene to Present

6. VOLUTA

Phylum	-	Mollusca
Class	-	Gastropoda
Order	-	Prosobranchea
Sub order	-	Pectinibranchea

A. Morphology

i) Shell thick, ovate, with short spire.

ii) Last whorl very large; on the posterior part are nodules or spines which are continued anteriorly as transverse (or axial) ribs.

iii) Aperture elongate, narrow, with an angular channel at the posterior end ; broad, and deeply notched at the anterior end.

iv) Outer lip thick; inner lip with thin callus. Columella with several transverse, only slightly oblique folds.

B. Age: Eocene to Present

7. CONUS

Phylum	-	Mollusca
Class	-	Gastropoda
Order	-	Prosobranchea
Sub order	-	Pectinibranchea

A. Morphology

i) Shell conical, smooth, the last whorl enveloping the preceding whorls.

ii) Spire short, conical, with many whorls.

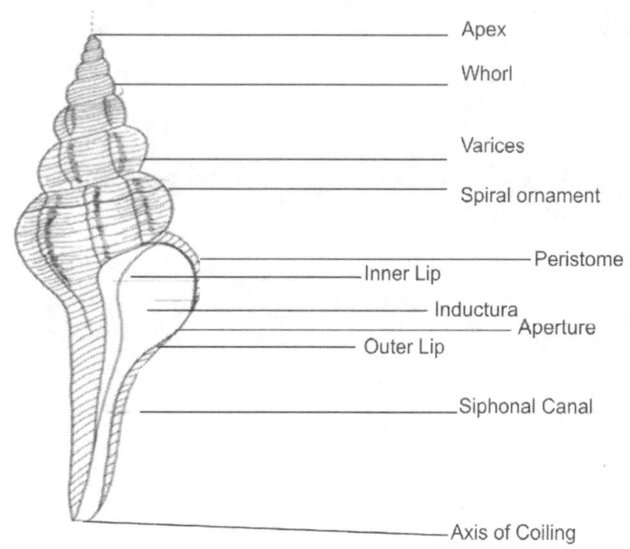

Gastropoda Morphology

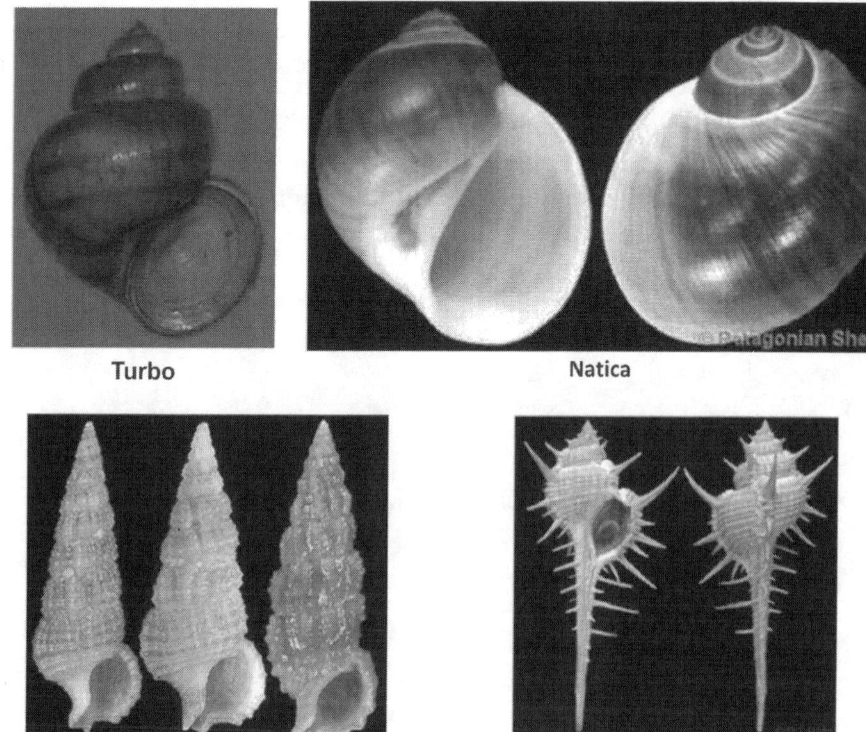

Turbo **Natica**

Cerethium **Murex**

(*See colour version on page 228*)

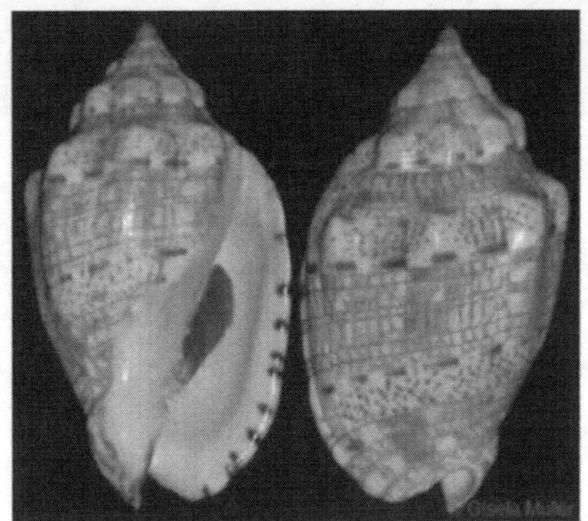

Voluta

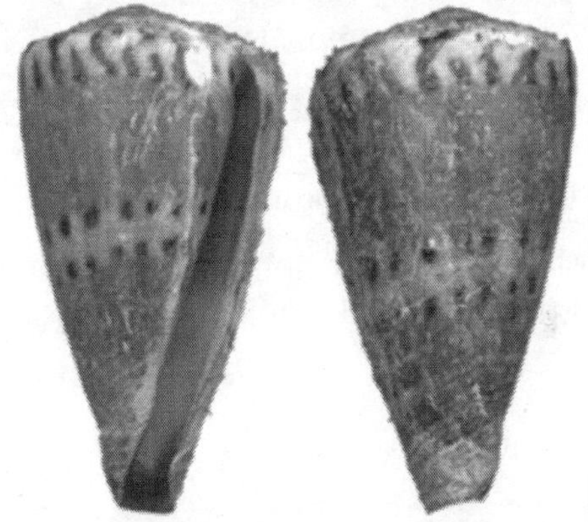

Conus

Planorbis

(*See colour version on page 228*)

iii) Aperture long, narrow, straight, with parallel or sub-parallel borders, ending anteriorly in a truncated canal.

iv) Outer lip thin, simple, no folds or teeth, notched at suture.

v) Columella straight, smooth.

vi) Operculum horny, much smaller than the aperture.

vii) The outer shell is thick, but the inner parts of the whorls become resorbed and very thin.

B. Age: Upper Cretaceous to Present

8. PLANORBIS

Phylum	-	Mollusca
Class	-	Gastropoda
Order	-	Pulmonata

A. Morphology

i) Shell sinistral, horny, whorls numerous

ii) Characteristic plane spiral is the only unique character in Planorbis. All Gastropods posses helical spiral

iii) Aperture oblique; peristome simple, sharp

B. Age: Jurassic to Present

III. Class Cephalopoda

The organism comes under class Cephalopoda are more advanced and well organized mollusks as compared to earlier ones. Nautilus and Cattle fish are the living example of cephalopods. Extinct varieties are belemnites, ammonites and goniatites. Cephalopods are mainly marine organism, great majority active free swimmers like modern nautilus. Some are benthonic like octopus. Morphologically they are spiral univalve shell. They are more complex and can be distinguished essentially by the following characters: i) Interior divided up in to chambers, whereas gastropods shell lacks internal divisions. ii) Shell exhibit bilateral symmetry and coiled plano-spiral, unlike the helicoids spiral of gastropods. There are exceptions- some have straight shell like belemnites. Based on morphological characters class cephalopoda can be divided into three orders:

i) Order Nautiloidea

ii) Order Ammonoidea

iii) Order Dibranchia

A. Order Nautiloidea

Nautiloid cephalopods are entirely marine. They were at their climax in Palaeozoic period, but presently only one species Nautilus has survived. Nautilus is characterized by simple nautiloid suture line.

1. NAUTILUS

Phylum	-	Mollusca
Class	-	Cephalopoda
Order	-	Tetrabranchia
Sub order	-	Nautiloidea

A. Morphology

i) Shell globose, spiral, plane spiral, whorls few completely embracing to each other.

ii) Umbilicus small or absent. Body chamber much larger than the preceding one,

iii) Aperture simple, with an external sinus.

iv) Septa concave, suture-lines more or less undulating. Siphuncle central, septal necks short and directed backwards. Surface of shell smooth or ornamented with striea or ribs.

B. Age: Trias to Present

2. ORTHOCERAS

Phylum	-	Mollusca
Class	-	Cephelopoda
Order	-	Tetrabranchia
Sub order	-	Nautiloidea

A. Morphology

i) Shell straight or occasionally slightly curved elongate -conical;

ii) Transverse section usually circular. Septa concave; suture-line straight.

iii) Body-chamber large; aperture not contracted or produced into lobes.

iv) Siphuncle cylindrical, without internal calcareous deposits; usually central, but sometimes sub-central or ex-central.

v) Ornamentation variable. Orthoceras, includes numerous species which have been grouped into more restricted genera based mainly on the character of the ornamentation.

Age: Upper Cambrian to Triassic

B. Order Ammonoidea

The Ammonoids are best index fossils as the organisms survived between Devonian to Cretaceous. etc. The shell is generally coiled into a plane spiral, and the suture-lines show complex patterns. The siphuncle is at the margin of the shell generally at the outer margin, but occasionally present at inner margin, it is usually more slender than in the Nautiloids and does not contain internal calcareous deposits.The form of the suture-lines varies considerably in different genera and is of great importance for systematic purposes. The portions of the suture-line which are convex towards the mouth of the shell are termed saddles while the intervening concave portions are known as the lobes. In many forms the lobes and saddles exhibit secondary folding which may be slight, producing merely a denticulate pattern, or may be deep and provided with other smaller folding, giving a foliaceous appearance to the suture. The lobes and saddles are nearly always similar on the two sides of the shell: commonly there is first the external lobe at the external margin, then the superior and inferior (or first and second) lateral lobes on the sides of the whorl and near the inner margin other lobes known as auxiliary lobes may occur. The evolution of ammonoids is best exhibited by the development of suture line (lobes and saddles).

Description of fossils

1.GONIATITE

Phylum	-	Mollusca
Class	-	Cephalopoda
Order	-	Tetrabranchia
Sub order	-	Ammonoidea

Evolution of Ammonitic suture line with age.

(*See colour version on page 229*)

A. Morphology

i) Shell smooth or with striations, whorls wide, younger whorl embracing older one.

ii) Umbilicus small or covered by whorls.

iii) Septal neck short projected backward but small part directed forward.

iv) Primitive Goniatite exhibits characteristic goniatitic suture (rounded saddles projecting outer side and angular lobes projecting inside).

v) In advanced variety the external lobes exhibit denticulations like ceretites.

B. Age: Carboniferous Age (Index Fossil)

2. CERETITES

Phylum	-	Mollusca
Class	-	Cephalopoda
Order	-	Tetrabranchia
Sub order	-	Ammonoidea

A. Morphology

i) Shell is discoidal

ii) Ribs containing tubercles present near umbilicus and external margin

iii) Umbilicus slightly large. Body chamber short

iv) Saddles are rounded and lobes are denticulated

B. Geological Age

Triassic (Index Fossils).

3. SCAPHITES

Phylum	-	Mollusca
Class	-	Cephalopoda
Order	-	Tetrabranchia
Sub order	-	Ammonoidea

A. Morphology

i) Shell coiled in plane spiral. whorls started from centre and as it proceed further, covers the early whorl.

ii) Last whorl is free from spiral, and then re curved in the form of a hook.

iii) Surface is ornamented with bifurcated ribs

iv) It contains tubercles

v) Suture lines are much divided

Age: Upper Cretaceous. (Index Fossil)

4. ACANTHOCERAS

Phylum	-	Mollusca
Class	-	Cephalopoda
Order	-	Tetrabranchia
Sub order	-	Ammonoidea

i) Whorl thick, Umbilicus large

ii) Ribs are simple or bifurcated

iii) Rows of tubercules at the side and margin

iv) External margin is broad with a median row of tubercles

v) Saddle broad

B. Geological Age

Cretaceous/ Lower Chalk (Extinct).

5. PERISPHINCTES

Phylum	-	Mollusca
Class	-	Cephalopoda
Order	-	Tetrabranchia
Sub order	-	Ammonoidea

A. Morphology

i) Shell discoidal, external margin rounded

ii) Umbilicus large, Ribs straight, continuous

iii) Bifurcating near the external border

iv) Constrictions are present at intervals on the whorl

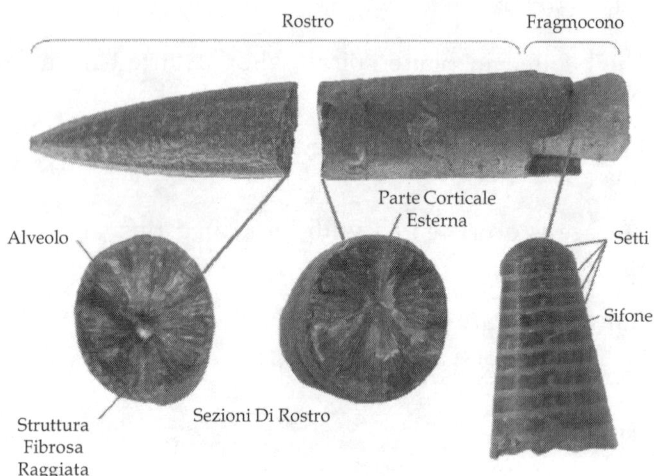

Balemnites

Orthoceras

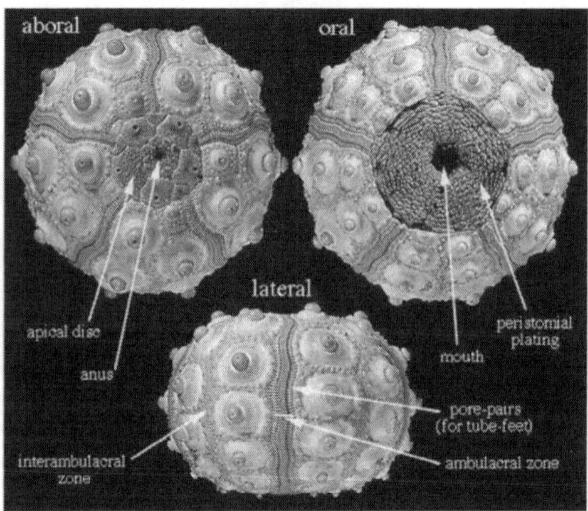

(*See colour version on page 230*)

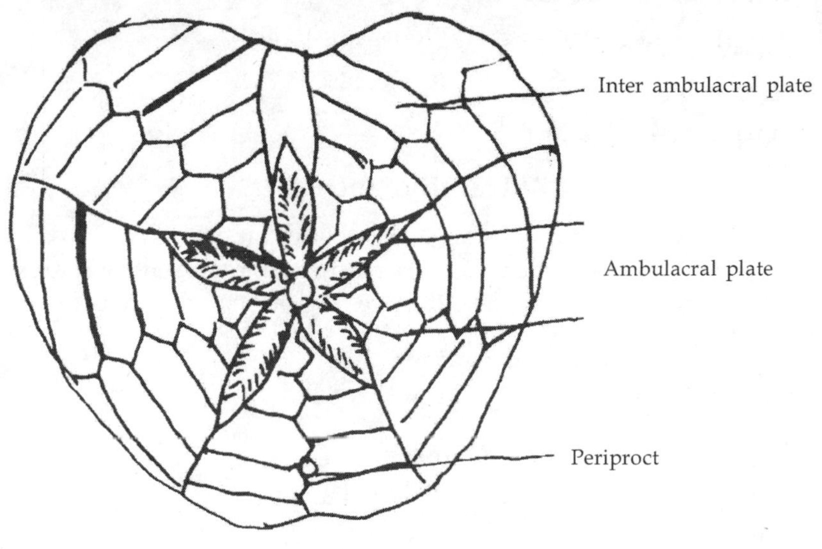
Inter ambulacral plate
Ambulacral plate
Periproct

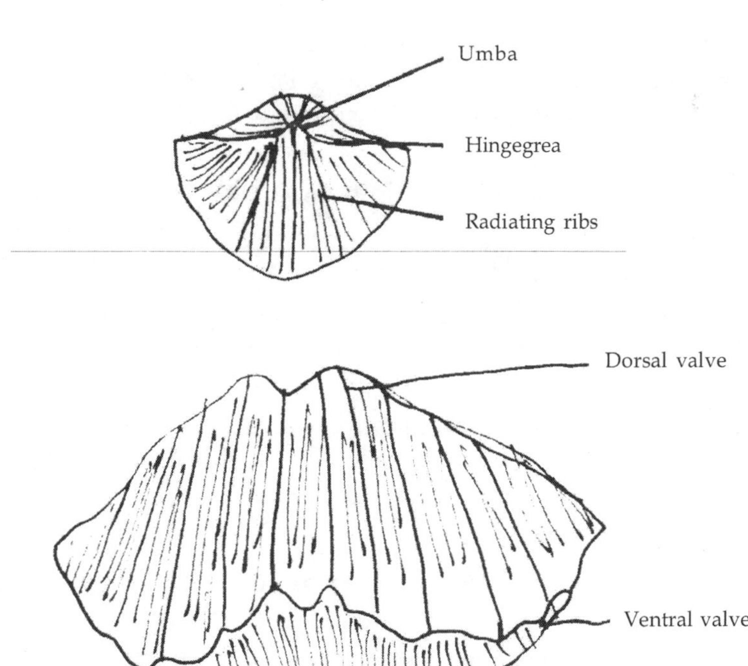
Umba
Hingegrea
Radiating ribs
Dorsal valve
Ventral valve

v) Suture line is much divided

vi) External and superior lateral lobes are large

B. Geological Age

Upper Jurassic (Extinct)

IV. Class Dibranchia

The Dibranchia are much less geologically important than fossils of nautiloidea and Ammonoidea. Class Dibrachia are represented at present day by the cuttle fishes, octopus etc. In geological record only Belemnites are present as fossil form.

1. BELEMNITES

Phylum	-	Mollusca
Class	-	Cephalopoda
Order	-	Dibranchia
Sub order	-	Decapoda

A. Morphology

i) The shell consists of three parts the guard or rostrum (a), the phragmocone (b), and the pro-ostracum (c).

ii) The guard is solid and is commonly preserved than the other parts; it varies in shape and size, (cylindrical, club-shaped, fusiform, conical, etc).

iii) The original form seems to have been a short cone. The end which was directed away from the mouth.

iv) The guard varies in length from one to fifteen inches.

v) When sliced transversely or longitudinally it is seen to be formed of a number of layers (growth-layers) arranged concentrically around an axial line.

vi) Each layer is formed of minute prisms of calcite, which are placed perpendicular to the axial line.

vii) The surface of the guard is smooth, or it may be granular, or furnished with ramifying vascular impressions.

viii) The phragmocone is a hollow cone, part of which fits into the alveolus at the broad end of the guard.

B. Geological Age

Lower Jurassic to the Upper Cretaceous.

5.2.5 Phylum Echinoidea

The echinoids or sea-urchins are globular, heart shaped, or disc shaped body, ornamented with spines. The mouth is on the inferior surface, and is either central or in front of the centre. The anus is at the posterior side of test. In the regular echinoids both anus and mouth are central being placed at opposite poles of the test; in the irregular echinoids the anus is always, and the mouth often eccentric.

The test is composed of three parts: a small patch of plates placed at the summit, known as the apical disc or apical system; the main part of the test termed the corona-, and the part between the mouth and the lower margin of the corona, which usually bears plates and is known as the peristome.

In the regular Echinoids, the anus is placed within the apical disc, which then consists of the following parts. Near the centre is the anus (a), which is surrounded by a membrane bearing small plates and known as the periproct (p). The periproct is encircled by a ring formed of ten plates, five are called genital (g) and five ocular (o). The genital plates form the inner part of the ring; they are often more or less hexagonal in outline, and are usually provided with a perforation which serves as the opening for the genital ducts. Outside the genital plates and alternating with them are the ocular plates; these are smaller than the genital and usually triangular or pentagonal in shape.

Description of Echinod fossils

1. CIDARIS

Phylum	-	Echinoderma
Sub Phylum	-	Eleutherozoa
Class	-	Echinoidea
Order	-	Regularia
Sub order	-	Endobrachiata

A. Morphology

i) Test spheroidal, the summit and base equally flattened. Apical disc very large, rarely preserved fossil,

ii) Ocular plates large and exsert. Ambulacra narrow, flexuous or nearly straight;

iii) Plates numerous, simple, all similar in form, pores unigeminal; between the rows of pores are vertical rows of small tubercles and granules.

iv) Interambulacra wide, plates large, each with a primary tubercle which is perforated, and may be crenulated or smooth; areola large, surrounded by secondary tubercles, beyond which may be granules.

v) Peristome large, without incisions, its membrane covered with plates. Spines large, of various forms, generally ornamented with rows of granules.

B. Geological Age:

Jurassic to Present

2. MICRASTER

Phylum	-	Echinoderma
Sub Phylum	-	Eleutherozoa
Class	-	Echinoidea
Order	-	Irregularia
Sub order	-	Spatangina

A. Morphology

i) Test heart-shaped or oval,truncated behind

ii) Apical disc small, excentric; madreporic plate extending to the centre; posterior genital absent

iii) Ambulacra sub-petaloid, placed in sunken areas, the sub-petaloid parts of the two anterior lateral longer than those of the two posterior lateral; pores unigeminal

iv) The anterior unpaired ambulacrum in a deep groove, with its pores circular. Interambulacra with large plates; tubercles small, perforate and crenulate. Fascicle below the anus

v) Peristorne near the anterior border, with a projecting lip (labrum). Periproct on the upper part of the posterior end.

B. Geological Age

Cretaceous to Miocene

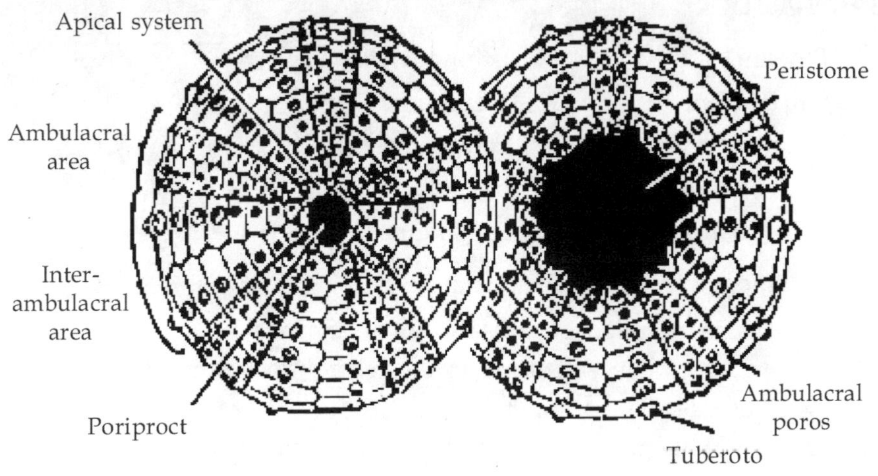

Micraster Fossil Sea - Urchin, Upper Chalk

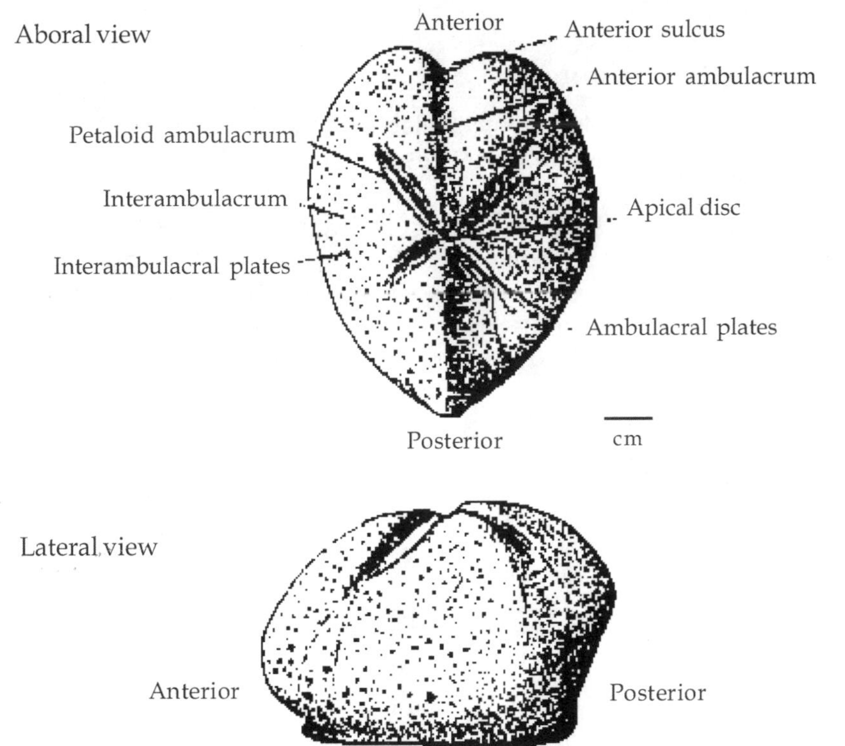

3. HEMIASTER

Phylum	-	Echinoderma
Sub Phylum	-	Eleutherozoa
Class	-	Echinoidea
Order	-	Irregularia
Sub order		Spatangina

A. Morphology

i) Test is heart shape, Apical disc small, excentric.

ii) Medriporic plate extending to the centre. Posterior genital absent.

iii) Ambulacral plate sub petaloid.

iv) Inter ambulacral with large plates, Tubercles small.

v) A peripetalous fasciole only present. Pores slit-like in the petaloid parts of the ambulacra, except in the anterior ambulacrum.

B. Geological Age

Cretaceous to Present

5.2.6 Phylum Arthropoda

Phylum Arthropoda to which trilobites belong, consists of organisms which alone among invertebrates, posses jointed limbs used as walking, swimming, food gathering and feeding organs. Arthropods have an external skeleton of chitin, an organic substance, hardened by calcium carbonate or calcium phosphate. Phylum Arthropoda are divided into five classes in which only one class Crustacea contains aquatic animals and are represented as fossils under Trilobita as a sub class. The body of a Trilobite is divided into three parts Head, Thorax and Pygidium.

Sub Class Trilobita

Description of fossils

1. CALYMENE

Classification

Phylum	-	Arthropoda
Class	-	Crustacea
Sub Class	-	Trilobita
Order	-	Proparia

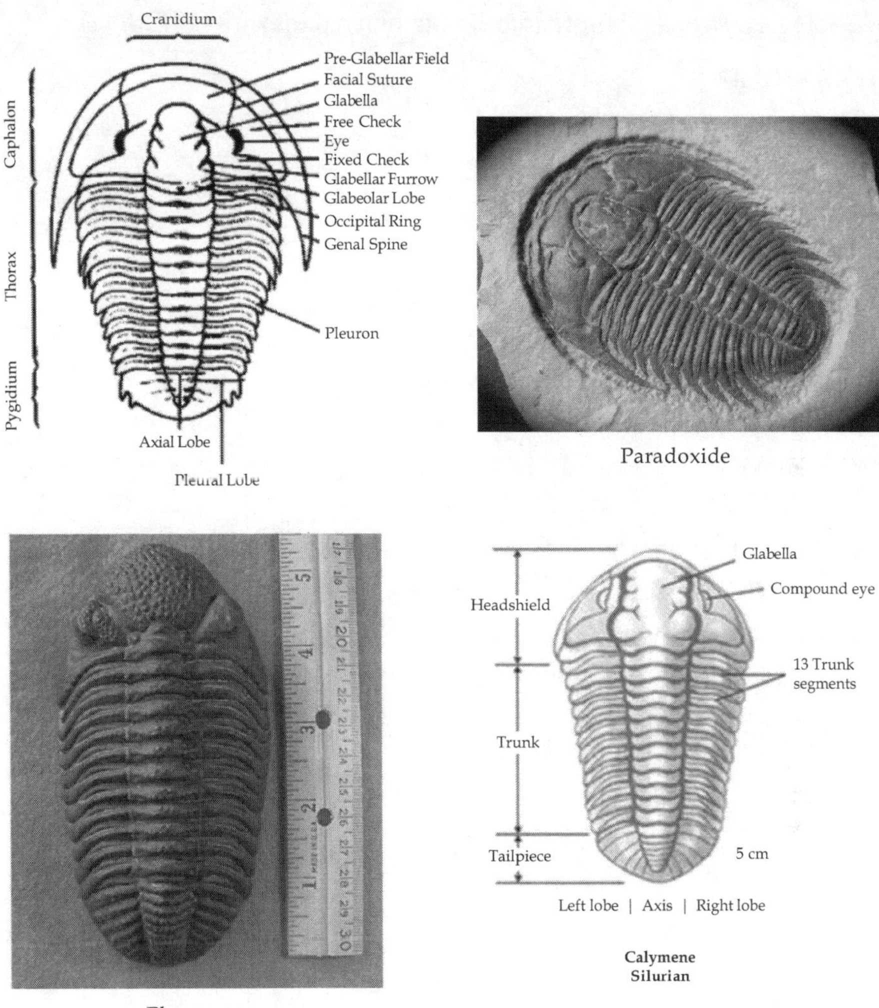

Paradoxide

Phacops

Calymene
Silurian

(*See colour version on page 231*)

A. Morphology

1. Head shield is semicircular and developed
2. Genal angle generally rounded occasionally angular
3. Glabella furrow is clear. Eyes are small but clear
4. Thorax segments are 13 in number
5. Pygidium is semicircular in which 5 to 8 segments fused.

B. Geological Age: Silurian to middle Devonian. Index fossil (Extinct)

2) PARADOXIDE

Classification

Phylum	-	Arthropoda
Class	-	Crustacea
Subclass	-	Trilobita
Order	-	Proparia

A. Morphology

1. Extremely large trilobite growing up to 60cm in length.
2. Thorax segment are 21 in number
3. Cephalon also contains a pair of long spines known as genal spines. It is pointing towards pygidium.
4. The last 2 to 8 segments fused and form pygidium
5. Pygidium is semicircular in which 5 to 8 segments fused

B. Geological Age-Middle Cambrian (Index Fossil)

3. PHACOPS

Phylum	-	Arthropoda
Class	-	Crustacea
Subclass	-	Trilobita
Order	-	Proparia

A. Morphology

i) Head shield more or less semicircular.
ii) Glabella prominent, broad in front with 3-4 furrows.
iii) Facial suture is present in lateral borders of cheeks in front of genal angle.
iv) Eyes are compound large appearing very distinct on glabella.
v) Thorax having 11 segments. Pygidium small.

B. Geological Age

Ordovician to Devonian (Index fossil).

5.2.7 Description of Plant Fossils

Appearance of Plant fossils in earth's history was best exhibited in Carboniferous and Permian period also known Gondwana period.

Gondwand time was very favorable for luxuriant growth of plants and all the coal deposits of the world belong to Gondwana period.

Description of fossils

1. GLOSSOPTERIS

Classification

Phylum	-	Spermatophyta
Class	-	Gymnosperms
Order	-	Cycadofilicales
Name	-	Glossopteris

A. Morphology

i) The leaves of glossopteris are simple and tongue shaped.

ii) The length of leaf is few centimeter to decimeter.

iii) Midrib is prominent and shows reticulate venation.

iv) It occurs in the Lower Gondwana sequence the name given is Glossopteris flora.

B. Geological Age - Lower Gondwana (Permian to Jurassic)

2. GANGAMOPTERIS

Classification

Phylum	-	Spermatophyta
Class	-	Gymnosperms
Order	-	Cycadofilicales
Name	-	Gangamopteris

A. Morphology

i) The Gangamopteris leaf is simple tongue shaped.

ii) The length of leaf is few centimeter to decimeters.

iii) It shows parallel venation. There is no midrib in Glossopteris.

iv) It is characteristic plant fossil of lower Gondwana.

B. Geological Age - Lower Gondwana (Permian to Triassic)

3. VERTIBRARIA

Classification

Phylum	-	Spermatophyta
Class	-	Gymnosperms
Order	-	Cycadofilicales
Name	-	Vertibraria

A. Morphology

i) Vertibraria is a rizome (underground stem) of Glossopteris

ii) It contain rectangular Blocks is longitudinal and transverse direction

iii) The rezome contains nodes and Internodes.

B. Geological Age - Lower Gondwana (Permian to Jurassic)

4. PTILOPHYLLUM

Classification

Phylum	-	Spermatophyta
Class	-	Gymnosperm
Order	-	Cycadales
Name	-	Ptilophyllum

A. Morphology

i) The leaf is compound and leaflets are sickle shaped.

ii) The small leaflets are called Pinnate.

iii) Pinnate shows parallel venation.

iv) All pinnets arranged one above other in a central rod like structure called Rachis.

B. Geological Age - Upper Gondwana (Triassic to Cretaceous)

Glossopteris characterized by prominent midrib Lower Gondwana.

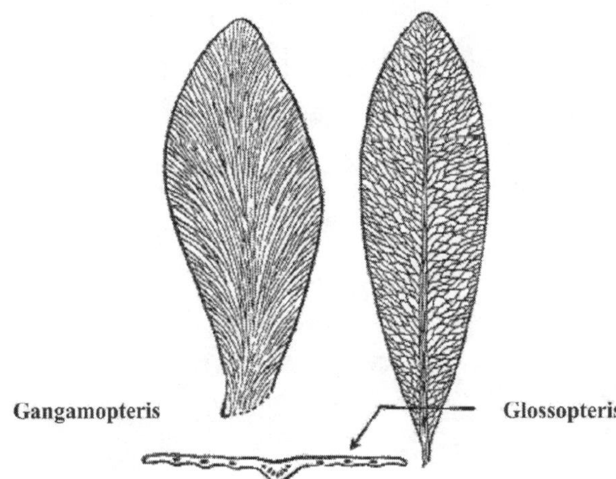

Vertibraria (Fern Rizome) Lower Gondwana

Compound leaves of Ptilophyllum Upper Gondwana

(*See colour version on page 232*)

Chapter 6

Stratigraphy

6.1 Introduction

Stratigraphy is the branch of geology which deals with the study of rock layers (strata) and layering (stratification) in their chronological order. It is primarily used in the study of sedimentary and layered volcanic rocks in order to interpret geological history and age. The term historical geology is also used as a synonym of stratigraphy but in a wider sense, it deals with the whole sequence of events that make up the history of the earth from its origin and age relations of rock strata, evolution through its life form, and structures. The scope of stratigraphy is very wide; it involves description and classification of layered sedimentary rocks and also of igneous & metamorphic rocks. The study involves correlation between different areas like lithological characters, fossil content, geophysical and geochemical properties including absolute radiometric dating.

6.2 Principles of Stratigraphy

Stratigraphy is based on the following few principles:

6.2.1 Law of superposition

In a sequence of horizontal sedimentary rock layers, the younger layers are on the top and the older layers are at the bottom. To put it another way, in a normal sequence, each bed is younger than the bed beneath it. Geologists today use it as the basis for a bed-by-bed analysis of geological history in a given area. This is important to stratigraphic dating, which assumes that an object cannot be older than the materials of which it is composed. This law known as "Law of Superposition" was first proposed by the scientist Nicolas Steno in the 17th century.

6.2.2 Law of uniformitarianism

The "law of Uniformitarianism" was proposed by Charles Lyell and James Hutton which states that "Present is the key to the past". The principle states that the same natural laws and processes that operate in the universe now have always operated in the past and apply everywhere in the universe. By this law a geologist can learn how ancient sedimentary rocks were formed on the basis of studies on modern sediments.

6.2.3 Rock units/Formation/Supergroup

The stratigraphic study also follows a systematic nomenclature to describe rock units. The smallest unit of a stratified sequence is known as **Bed**, and it consists of single stratum separating it from other layers above and below. A bed is not given a formal name unless it is specific, like coal bed or marker bed of stratigraphic significance. Sequence of beds of same composition deposited and forms a thick sequence of sedimentary rock, is called **Formation**. Two or more formations in succession form the **Group**; likewise more than one Group occurring together in a continuous manner in the given area forms the Supergroup.

6.2.4 Sedimentary facies

Sedimentary facies are bodies of sediment recognizably different from adjacent sediment deposited in a different depositional environment. Generally, facies are distinguished by the aspect of the rock or sediment it is being studied. Thus, facies based on petrological characters such as grain size and mineralogy are called *lithofacies*, whereas facies based on fossil content are called biofacies. Sedimentary facies are the result of variations in current velocity and other variables in the basin of deposition (which make environment of deposition).

6.2.5 Stratigraphic correlation

Geologic study concerned with establishing geochronological relationships between different areas, based on geologic investigations of many local successions is called stratigraphic correlation. The correlation aims at matching the rock formations of distant areas deposited at the same stage of the earth's evolutinary history. The two formations thus correlated should have been deposited at the same time, thus they are synchronous in their origin. These formations deposited at the same stage of the earth's evolutionary history are termed homotaxial.

6.3 Stratigraphic Classification

Correlation of rock formations of different region forms an important theme of stratigraphy. The correlation aims at matching the rock formation of distant areas deposited at the same stage of the earth's evolutionary history. **IUGS (International Union of Geological Sciences)** has given following terminologies in order to bring uniformity in stratigraphic classification worldwide: Lithostratigraphic, Chronostratigraphic and biostratigraphic Units.

6.3.1 Lithostratigraphic unit & correlation

This stratigraphic correlation is mainly based primarily on lithologic criteria which is easily recognizable in the field. These units are characterized by regional names and used as geological, structural and economic description of that region. The lithostratigraphic correlations are made by the following methods:

i) Continuity of lithological units and their sequence in that order (continuity of litho-contacts)

ii) Lithologic similarity

iii) Well logs

iv) Structural characteristics.

The following are the formal terminologies used in lithostratigraphic classification:

I. **Formation** : It is a fundamental unit of lithostratigraphic classification. A Formation represents rock succession deposited under a relatively uniform physiochemical condition.

II. **Member** : Member denotes a part of formation, having some characteristic fossil assemblages. It is smaller unit than formation but mapable.

III. **Group** : It denotes the thick succession of rock spread over a large area.

IV. **Supergroup** : An association of mutually related Groups is called Supergroup.

6.3.2 Chronostratigraphic unit & correlation

A rock unit is characterized by its age of deposition which is determined by geochronological methods is known as chronostratigraphic unit. It has identical names similar equivalent ranks with geochronological units.

I. **Erathem** : Succession of rock formation deposited during an Era.

II. **System** : Succession of rock formation deposited during a period. Example: Tertiary System.

III. **Series** : A Series is a part of the system which was deposited during an Epoch. Example- Miocene series.

IV. **Stage** : A stage is a part of series deposited during a geological age. Example: Panonian Stage.

6.3.3 Biostratigraphic units and correlation

The paleontological evidences are considered as reliable indicators of age and evolution of life. These record constitutes biostratigraphic units. Zones are considered as separate kinds of stratigraphic unit called Biostratigraphic unit. A zone is confined to certain smaller biogeographical areas. **Biostratigraphic zones may be defined by** :

1. Total stratigraphic range of one species;
2. A range in which one species of fossil occurs in abundance.

6.4 Physiographic and Tectonic divisions of India

Indian subcontinent shows distinct physiography in different parts which show different tectonic framework; which help in studying each unit separately. The Himalayan mountain range is result of relatively recent tectonic activity that India plate experienced. We have broad three tectonic units which coincide with the three physiographic divisions :

1. **Peninsula**,
2. **Extra Peninsula** and
3. **Indo-Gangetic Alluvial Plains**.

The Peninsula exhibits a geology that has not experienced any major deformation since the beginning of the Phanerozoic eon. The earliest rocks have been severely deformed during the Archaean and Proterozoic times. Eventually when they got stabilized, they formed basement for other rocks to be laid over them. We will study these earliest rocks under the heading **Precambrian Basement**. Precambrian Basement rocks are widespread in southern and south-eastern India and also in Aravalli Range of mountains in Rajasthan.

Although there were phases of deformation in some parts during the Proterozoic, in some other places there had been quiet depositional activities during this time and these rocks still lay nearly flat covering large areas. We group such successions under **Proterozoic Cover**. Large basins with Proterozoic Cover rocks are exposed in north-central India and parts of Andhra Pradesh, Karnataka, Madhya Pradesh, Chhattisgarh, Jharkhand and Maharastra.

There were only occasional rock-building activities in Indian Peninsula after the close of the Precambrian. The southern and central part remained above sea-level. Therefore, marine depositional activities are normally not there except at Umaria, Manendragarh and Daltonganj along the Narmada Son graben rift zone. The Palaeozoic of Indian Peninsula are represented by river deposits laid down in wide grabens. The deposition by ancient river systems continued until the end of the Mesozoic. These rocks are grouped under the **Gondwana Supergroup**, which occur in linear belts (graben) along rivers such as Damodar, Godavari, Mahanadi, Wardha and also in Satpura Hills of Madhya Pradesh (Narmada-Son valley).

Apart from the river deposits, the Mesozoic also witnessed a unique event in Indian history. Close to the end of Mesozoic, vast areas of western India were covered by lava flows identified as Deccan flood basalt (Deccan Traps). Similar lava flows were laid down in eastern India in Rajmahal Hills. Sedimentary deposition was confined to coastal areas which were occasionally flooded by the transgressing sea. **Mesozoic** and **Cenozoic** marine formations are found in coastal areas such as Kutch, Saurashtra, Jaisalmer, Assam-Arakan belt and Cauvery Basin.

The Extra Peninsula because of its uniqueness has been treated separately as Himalayan Geology. The Indo-Gangetic Alluvial Plains are mostly covered with recently deposited alluvium carried by two major river systems -- the Ganga and the Indus.

6.5 Geological Time Scale

The **Geological Time Scale (GTS)** is a system of chronological measurement that relates stratigraphy to time, and is used by geologists, paleontologists, and other Earth scientists to describe the timing and relationships between events that have occurred throughout Earth's history.

Phanerozoic Eon

- Cenozoic Era
  - Quaternary Period
    - Holocene Epoch (.01)
    - Pleistocene Epoch. (1.8)
  - Tertiary Period
  - Neogene sub Period
    - Pliocene Epoch (5.3)
    - Miocene Epoch (23.8)
  - Palaeogene sub period
    - Oligocene Epoch (33.7)

Eocene Epoch (54.8)
Palaeocene Epoch (65.0)
Mesozoic Era
Cretaceous Period (142.0)
Jurassic Period (205.7)
Triassic Period (248.2)
Palaeozoic Era
Permian Period (290.0)
Carboniferous Period (354.0)
Devonian Period (417.0)
Silurian Period (443.0)
Ordovician Period (495.0)
Cambrian Period (545.0)
Precambrian Eon
Proterozoic Era (2500)
Archaean Era (3500- 4500?)

Geological Time Scale (Figures in parentheses give age in Million years)

6.6 Indian Stratigraphy

6.6.1 Dharwar Supergroup

Archaeans of south India is also known as Dharwar Supergroup. It was first studied by R Bruce Foot (1888). He gave the name Dharwar-System for schists, gneisses and granitic rocks, which are best developed in Mysore and Southern Bombay. The Dharwar rocks classified into two parts by Smeeth (1916), the lower referred to as Hornblendic Division and the upper as Chloritic Division. Ramarao (1936) proposed three fold classification: Lower, Middle and Upper Dharwar. Lower Hornblendic Division and upper chloritic division has been divided into middle and lower Dharwar division. Lower Dharwar (Bababudan Group) is made up of highly crushed micaceous quartzschists and gneisses, formed due to metamorphism of pre-existing acid volcanic rocks. Middle Dharwar (Chitradurga Group) is composed of metamorphosed igneous and sedimentary rocks. The upper part of the sub division consists of metamorphosed granitic rocks with gneissose structure. The lower part consists of micaceous gneiss, schistonse conglomerates, banded ironstones etc. Upper Dharwar is composes of mainly sedimentary rocks such as conglomerates, calcareous, cherty and ferrugenous silts and clays. Radhakrishanan (1964) studied older schist belt in Dharwar province and given name Sargur Schist and thus the stratigraphic succession can be written as follows:

Closepet Granites

Dharwar Supergrous

Rannibennur Group

Chitradurg Group

Bababudan Group

Peninsular Gneissic Complex

Sargur Schist

6.6.2 Proterozoic/Purana formations

The Purana rocks defined a large spread of Middle to Upper Proterozoic platform cover deposits, which are preserved in several isolated basins in Peninsular India. These sediments, despite their development in widely separated areas show some similarity in characters like unmetamorphosed nature, mostly unfossiliferous and very little tectonic deformation. Major Purana basins in India are Cuddapah, Vindhyan, Chhattisgarh, Indravati, and several satellite basins like Albaka, Khariyar, Sukma, Ampani and Keshkal etc.

6.6.3 Cuddapah Supergroup

The name Cuddapah is derived from the Cuddapah basin of Andhra Pradesh. The rocks are exposed in a crescent shaped 340 km long basin between Singareni coalfield in the north and the Nagari Hills in the south near Madras. The average width of basin is 145 km. The basin is spread approximately in 42000 sq. km. Kurnool is the youngest formation of Cuddapah basin and is confined to north western part of the basin. The Cuddapah sediments are underlain by gneiss and schists of Archaean age. The unconformity plain is popularly known as Eparchaean Unconformity. The maximum depth is 3000 to 4000 metres in the concave (Easter part of the basin) portion.

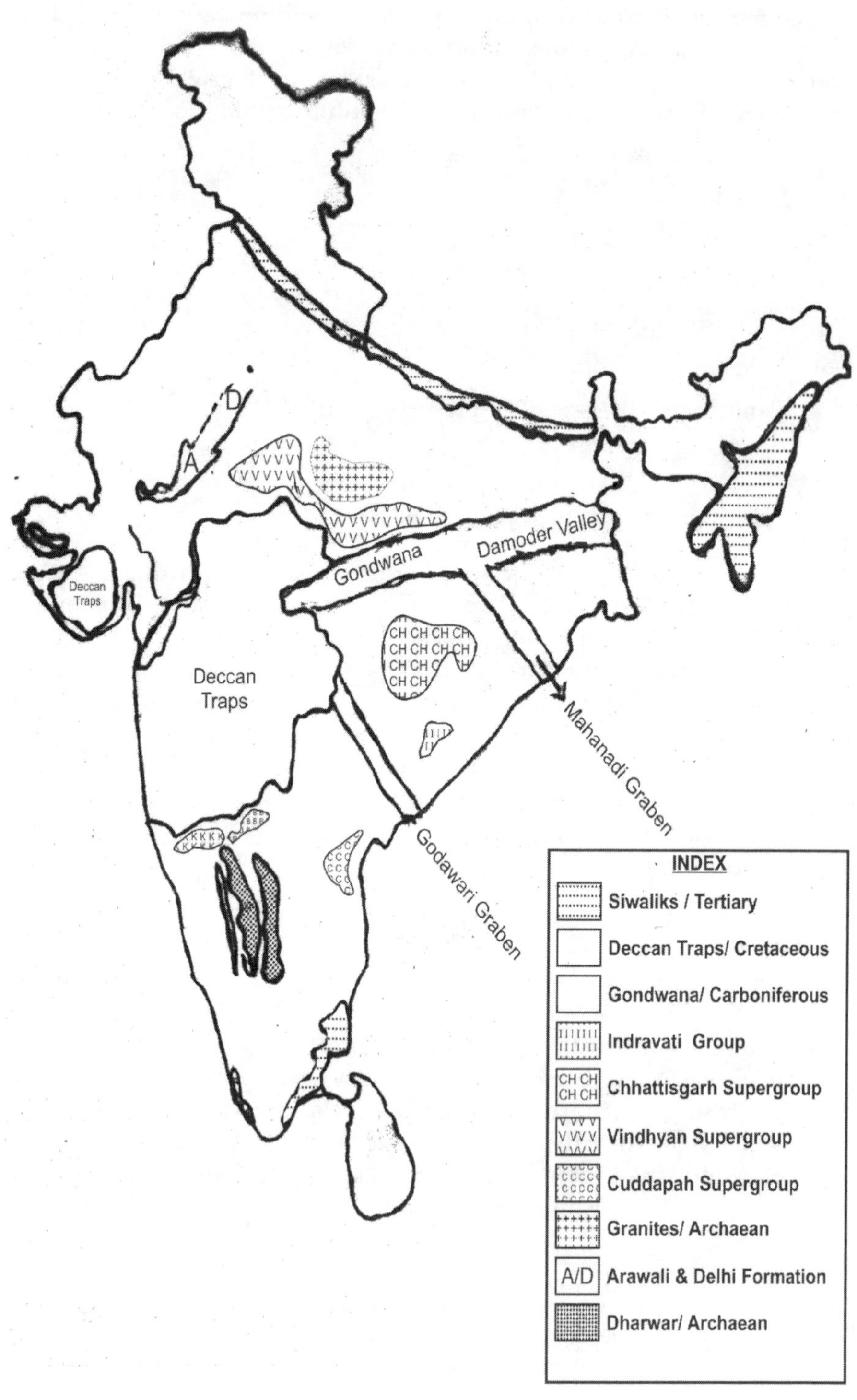
D
A
Deccan Traps
Gondwana
Damoder Valley
Deccan Traps
Mahanadi Graben
Godawari Graben
INDEX
Siwaliks / Tertiary
Deccan Traps/ Cretaceous
Gondwana/ Carboniferous
Indravati Group
Chhattisgarh Supergroup
Vindhyan Supergroup
Cuddapah Supergroup
Granites/ Archaean
A/D Arawali & Delhi Formation
Dharwar/ Archaean

The stratigraphic succession proposed by King (1822) is as under:

Upper	Kurnool Group	Sedimentary Rock
	---------------------------- Unconformity ----------------------	
	Kisthana Series (600m)	Srisalem Quartzite Kolamnala Shale Irlakonda Quartzite
	---------------------------- Unconformity ----------------------	
	Nallamlai Series (1000m)	Cumbum Shale Bairenkonda Quartzite
	---------------------------- Unconformity ----------------------	
Lower	Cheyair Series (3300M)	Tadpatri Shale Pulivendla Quartzite
	---------------------------- Unconformity ----------------------	
	Papagni Series (1400m)	Vempalli Shale Gulcheru Quartzite
	---------------------------- Unconformity ----------------------	
	Archaean	Crystalline basement of Dharwar Cration

6.6.4 Vindhyan Supergroup

The vindhyans show extensive development in central India and consist of arenaceous and calcareous rocks with very small argillaceous bands. Vindhyan rocks are developed characteristically in the Sonc vallcy and in certain parts in Rajasthan, M.P. and A.P. These rocks are unmetamorphosed and horizontally exposed. The stratigraphic succession is shown in the following:

		Upper Bhander Sandstone
		Sirbu shale
	Bhander Group	Lower Bhander Sandstone
		Bhander Limestone
Upper	Diamondiferous bed	Ganurgarh Shale
Vindyans		Upper Rewa Sand stone
	Rewa Group	Jhiri Shale
		Lower Rewa Sandstone
	Diamondiferous bed	Panna Shale
		Upper Kaimur Sandstone
	Kaimur Group	Bijaigarh Shale
		Lower Kaimur Sandstone
	------------------------- Unconformity --------------------	
Lower		Rohtas Formation
Vindhyans	Semri Group	Khenjua Formation
		Porcellinite Formation
		Basal Formation

Stratigraphic Succession of Vindhyan Group

The lower part of the system is made up of calcareous and argillaceous sediments deposited under a marine environment. The upper part is made up principally of fossils except few organic remains present in Suketshales of Lower Vindhyan Rocks of Rajasthan. The equivalents of Vindhyans are Bhima, Kaladgi (Maharastra) and Karnool Group (A.P.).

6.6.5 Chhattisgarh Supergroup

The upper Proterozoic sediments of Chhattisgarh Supergroup covers an area of about 33,000sq.Km. in central part of Chhattisgarh and Orissa. It rests un-conformably over Lower Precambrians.

Based on revised geological succession of Chhattisgarh Basin is divided into two Sub Basins western Hirri and eastern Bardwar Sub Basins (Mukherjee *et al.* 2014). The stratigraphic succession of Chhattisgarh Rocks are as follows:

Stratigraphic succession of the Chhattisgarh Supergroup. (Mukherjee et al., 2014)

Chhattisgarh basin			
Hirri sub-basin (covers ~25000 km^2		Baradwar sub-basin (covers ~8000 kin^2)	
Formation	Major lithology	Formation	Major lithology
Alluvium/ Laterite / Gondwana Supergroup /Intrusive rocks			
Raipur Group		Kharsiya Group	
Maniari (>300 m)	Gypsiferous purple shale and dolomite	Nandeli	Gypsiferous purple shale and dolomite
Hirri (150 in)	Stromatolitic dolomite and black shale	Sarnadih	Sandstone and conglomerate
Raipur Group			
Tarenga (>300 m)	Belha Member argillaceous dolomite Dagauri Member tuff (903 ± 8 Ma)[1] and Kusmi Member shale	Churtela	Purple shale and Sukhda tuff (1000 Ma)[2]
Chandi (670 m)	Stromatolitic limestone dolomite with Deodongar Member sandstone-shale	Chandi	Dolomite/stromatolitic limestone
Gunderdehi (>300 m)	Calcareous shale with Dotapar Member sandstone	Gunderdehi	Calcareous shale with stromatolitic limestone
Charmuria (260 in)	Flaggy limestone, shale with Sirpur Member tuff (?) and Ranidhar Member cherty limestone	Sarangarh	Flaggy limestone and shale
Chandarpur Group (130 m)		**Chandarpur Group**	
Kansapathar Chaporadih	Quartzs arenite Glauconitic sandstone/	Kansapathar (150 m) Chaporadih	Quartz arenite Glauconitic sandstone/silt
	silt stone and black shale	(Gomarda) (600 m)	stone and black shale
Lohardih	Subarkose with basal conglomerate	Lohardih (150 in)	Subarkose with basal conglomerate
Singhora Group			
		Saraipali (>300 m)	Shale, tuff (1405 Ma)[3] and Stromatolitic limestone with Bhalukona Member sst and basic intrusive (Diabase, 1420 Ma)[4]
		Rehatikhol (30-75 m)	Sandstone with conglomerate
(Older basement Rocks of Dongargarh Supergroup, Chilpi Group Iron ore Group, Amgaon Group)		Older metamorphic and basements (Rocks of Raigarh-Surguja metamorphic belt, Sonakhan Group, Sambalpur Granites)	

J. Earth System Science, 123 April 2014

Stratigraphic succession of Chhattisgarh Supergroup

The Chhattisgarh Basin contains about 2500 m thick sediments of orthoquartzite-carbonate pelite suit, deposited in multiple sedimentary cycles. The basin had also affected by minor felsic volcanic and pyroclastics. Each cycle starts with arenaceous facies and ends up with shale-limestone facies. The basin is divided into two sub basins western Hirri and eastern Barodwar sub basins separated by Sonakhan high.

6.6.6 Gondwana Group

After the deposition and uplift of the Vindhyan rocks during the Pre-cambrian era, the peninsula witnessed no further deposition of sediments for a long time. During the upper carboniferous period, however there commenced a new cycle of sedimentation in interconnected inland basins of fluviatile and lacustrine origin. This new phase of deposition of sediments continues up to the end of the Jurassic period. These inland sediments occupying a vast tract in the peninsula constitute the Gondwana Group or system (named after the Gond Kingdom of M.P.) where they were first studied by H.B. Medlicott in 1872. The Southern continents of the present day namely Australia, Antarctica, south America and India were during the Gondwana period united together to form one continuous stretch of land, known as Gondwana land.

Climatic condition

The deposition of Gondwana sediments commenced under glacial climatic condition. Then there prevailed a warm and humid climate during the rest of the Upper, Carbonifeous and the whole of the Permian. Throughout triassic, there prevailed a dry and arid climatic conditions. During Jurassic, again the country appears to have witnessed a more or less warm and humid cliamte. Each individual cycle commenced with the deposition of coarse sand. The Gondwana rocks were subjected to tectonic disturbances during the Mid Triassic, Jurassic and Post-Eocene periods leading to the development of a number of faults in them. They were also traversed by doleritic, Lamprophyre dykes and sills which in the Damodar valley area are said to be genetically related to the Rajmahal traps. On the basis of the paleontological, stratigraphical and lithological criteria, two classifications have been proposed for Gondwana rocks.

(a) The two-fold classification proposed by C.S. Fox, M.S. Krishnan etc. They have divided the Gondwana rocks on the basis of floral characteristics into Lower Gondwana characterised by Glossopteris flora and the Upper Gondwana sediments marked by the advent of the Ptilophyllum flora.

These two divisions are separated from each other by unconformity. This classification is as follows:

Subdivisions	Series	Stages	Ages
		Umia	Low. Cretaceous
	Jabalpur		
		Jabalpur	Up. Jurassic
		Kota	Mid. Jurassic
Upper Gondwana	Rajmahal		
		Rajmahal	Low. Jurassic
		Maleri	Up. Triassic
	Mahadeva		
		Pachmarhi	Up. Triassic
	----------------	Unconformity	----------------
		Raniganj	Up. Permian
		Barren measures	Mid. Permian
	Damuda	Barakar	Low Permian
		Karharbari	Low.Permia
	Talchir	Rikba	Up. Carboniferous
		Talchir	Up. Carboniferous
		Boulder bed	Up. Carboniferous

-------------------------- Unconformity --------------------

Proterozoic Basement

6.6.7 Deccan Traps (Deccan Volcanics)

At the close of Mesozoic era, western, central and south India witnessed enormous lava flow through fissure eruption. The lava spread out far and wide in the form of horizontal sheets. Their step like or terraced appearance is suggestive of the name **Deccan Traps** to these volcanic formations. Trap is a Scandinavian word meaning steps. The areal extent of traps is about 500,000 sq. km. covering Bombay, Kathiyawad, Kutch, M.P. Central India. Because of their tendency to form flat-topped plateau-like features and their basaltic composition, they are also termed as **plateau basalts**. The eruption of volcanic materials was mostly through the fissures and the Deccan Traps are believed to be the result of fissure-type of eruption. But at a few places, like Girnar hills, etc. the eruptions were of the central-type showing differentiated rocks of various characters. The lava-flows generally occur in the form of horizontal sheets ranging in thickness from seven metres to as much as thirty metres. It has a maximum thickness of about 3300 meters near Bombay. The traps show well developed columnar jointing caused by tensile stresses, the result of contraction due to cooling. The columns are long and polygonal in shape. At places traps show amygdaloidal structure. The traps are essentially a basic rock of basaltic composition, and is a dark coloured or melanocratic rock. There are very little signs of differentiation but in some places like Pavagarh hills in Gujarat and the Girnar hills of Kathiawar differentiated rocks are also observed. According to Washington, the mineralogical composition of the trap-rocks are as follows:

Quartz	2 to 5%	Orthoclase	5 to 7%
Labradorite	40 to 50%	Pyroxene	30 to 40%
Iron-oxide	10 to 12%		

Stratigraphic Classification of Traps

The traps have been divided into three groups: lower, middle and upper with Lameta, Jabalpur and Bagh beds at the base. In other words, the stratigraphic classification of the deccan traps are as follows:

Upper Eocene Nummulitic limestone of Surat and Bharuch

-------------------------- Unconformity --------------------

Upper flow (450 m thick)	Flows with numerous inter-trappean sedimentary layers and volcanic ash beds.Exposed in parts of Bombay and Kathiawar. They contain remains of vertebrates and molluscan shells.
Middle flow (1200m thick)	Numerous thick ash beds, practically devoid of intertrappeans sedimentary layers.Exposed in Malwa and Madhya Pradesh.
Lower flow (150 m thick)	Associated with numerous fossiliferous inter-trappean beds and rare ash beds. Exposed in parts of M.P. and eastern areas.

-------------------------- Unconformity --------------------

Lameta formation/Bagh beds/Jabalpur formation/Older rocks.

6.6.8 Lameta Beds/ Infratrappeans

The fluvial and estuarine beds occur below deccan traps at Lameta Ghat, Jabalpur are known as Lameta beds. Equivalets of Lameta beds occur around Nagpur, Jabalpur, Godavari valley and western part of Narmada Valley. The thickness varies from 6 to 35 meters and the major rock types are limestone, sandstone and clays. Limestone is impure gritty and containing lumps of chert and jasper. Sandstone is greenish and clay occurs sandy red and green in colour. Lameta beds are shallow marine formations and contain small fragmentary fossils of Mollusca Physa bullinus (Physabullinus), fishes and dinosaurs.

6.6.9 Intertrappeans

The intertrappean beds, are fossiliferous flora formed by fluviatile and lacustrine deposits intercalated with deccan lava flow. They serve as a valuable guide to the history of animal and plants existed during that period. Intertrappean beds are 2 to 10 ft in thickness at places the thickness less than 1ft. There lateral extension ranges from 3 to 4 miles. They are distributed in the middle and upper Traps. The rock types present here are impure limestone, chert, pyroclastic materials etc. Main localities are Chhindwada, Nagpur, Dhar, Sagar and Jabalpur districts.

6.6.10 Siwalik system

The Siwalik Hills is a mountain range of the outer Himalayas also known as [Manak Parbat] in ancient times. Shiwalik; This hill range is about 2,400 km (1,500 mi) long enclosing an area that starts almost from the Indus and ends close to the Brahmaputra. The width of the Sivalik Hills varies from 10 to 50 km (6.2 to 31.1 mi), their average elevation is 1,500 to 2,000 m (4,900 to 6,600 ft).

Geologically, the Siwalik Hills belong to the tertiary deposits of the outer Himalayas. They are mainly composed of sandstone and conglomerate rock formations, which are the solidified detritus of the great mountain range to their north, but often poorly consolidated. They are the southernmost and geologically youngest east-west mountain chain of the Himalayas. They have many sub-ranges and extend west from Arunachal Pradesh through Bhutan to West-Bengal, and further westward through Nepal and Uttarakhand, continuing into Himachal Pradesh and Kashmir. The hills are cut through at wide intervals by numerous large rivers flowing south from the Himalayas.They are bounded on the south by a fault system called the Main Frontal Thrust, with steeper slopes on that side. Below this, the coarse alluvial Bhabar zone makes the transition to the nearly level plains.

At the end of lower miocene period, the second phase of the Himalayan Orogeny had initiated. The floor of the Tethys was up-heaved and the seas had disappeared leaving no basin. As the mountains were getting higher and higher, the numerous streams that had established their courses became more and more active by an increase in their gradient. Thick pile of weathered sediments were deposited at the foot of the mountains. Like a geosyncline basin, it subsided gradually due to deposition of sediment load. This fluviatile detritus was later subjected to mountain building forces during the third phase of upheaval of the Himalayas and gave rise to a series of hills called Shivalik hills. Thus the Siwalik system named after the Shivalik hill is developed all along the foot hills of the Himalayas Arc. The system is composed mainly of sandstone, conglomerate, silts and clays which were found under brakish and fresh water environment. An appreciable position of the Siwalik strata is devoid of fossils. Still then the mammalian fossils are of great significance in dividing this system into various sub-divisions.

The tertiary rock of India, are well developed in the extra-peninsular region. In the peninsula small patches of Tertiary rocks occur in Rajsthan and along the coastal tracts in Orissa, Madras, Gujarat etc.

All along the length of the Himalayas, tertiary rocks occur in the foot-hill region. Tertiary rocks ranging in age from middle-miocene to middle pleistocene are together known as Siwalik system which are particularly well developed in the western part of the Himalayan Region.

The tertiary succession in the Himalayan region is as follows:

	Upper	Boulder Conglomerate	Low to mid, pleistocene
		Pinjor formation	Up.pleistocene to
Siwalik System		Tatrot formation	Low.pleistocene
	Middle	Dhok Pathan formation	Low.pleistocene
		Nagri formation	Up.miocene
	Lower	Chinji formation	Mid. miocene
		Kamlial formation	Mid. miocene
------------ Unconformity ------------			
		Murrees	Lower to Middle
------------ Unconformity ------------			Miocene
Kasauli Formation (Upper Murrees)			

❑❑❑

Chapter 7

Structural Geology

7.1 Introduction

The earth's crust is made up of igneous, sedimentary and metamorphic rocks. Each variety of rocks exhibit some columnar joints in Igneous rock, characteristic structures like bedding in sedimentary rocks and fabrics like schistosity and gneissosity in metamorphic rocks.

As a result of tectonic activities in nature, rocks may suffer deformation, rupture, dislocation of beds (Fault) or other planar structures.. The whole study of such deformation is known as structural Geology. It has been observed that sedimentary rocks are most suitable for studying structures, because these rocks are deposited in quiet environment, and initially the beds arranged in horizontal position. As a result of stress due to kinematic movements/tectonism, the horizontal beds tend to deviate from their horizontal disposition to dipping position or get folded or faulted. Thus the rock develops various kinds of structures. There are following structures present in sedimentary rocks:

Describing orientation of geologic features with strike and dip

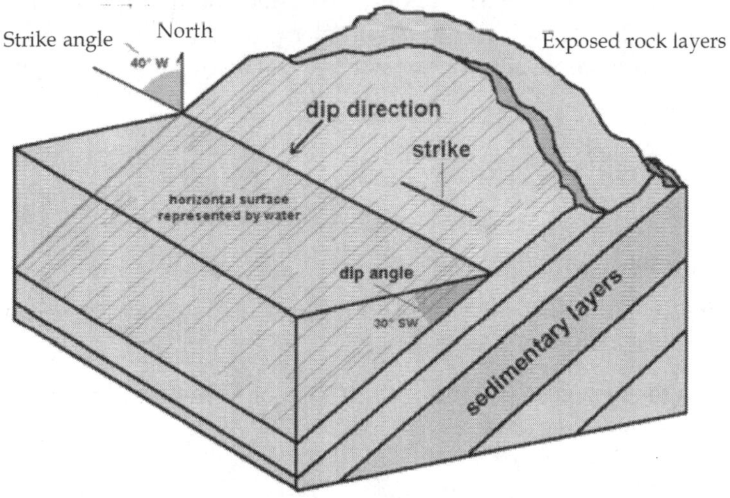

Fig. 7.1 Components of Dip and strike.

7.2. Primary Sedimentary structures

The structures that produced during the formation of sedimentary rock are known as Primary Sedimentary Structures. Main primary sedimentary structures are bedding, current bedding, ripple marks, mud cracks, rain prints etc.

7.3. Secondary or Deformational Structures

The structures produced due to result of tectonic movements are known as secondary or deformational structures. The secondary structures are of following types:

7.3.1 Dip

Sedimentary rocks initially deposited in horizontal position. Due to tectonic movements, these beds become tilted in any direction and at any angle, and are known as dips. Dip is generally described in terms of direction and amount. The geographical direction along which a bed shows maximum inclination is known as direction of dip. The line perpendicular to the dip is a strike line, which is described as attitude of horizontal line on the inclined/dipping surface given either full circle reading on the compass (90 degree, 180 degree) or quadrant measures (N40 degree W or S 40 degree E).

7.3.2 Folds

Folds are the wavy undulatory structures produced by the process of diastropism, particularly in deformed sedimentary rocks. Folds vary in size with the amplitude and wavelength of folded surface, varying from fraction of centimetre to even several hundreds metres. (in regional scale folding).

7.3.3 Elements of folds

Limb - A fold generally consists of two limbs. These are the sloping arms dipping in opposite side.

Crest - Generally measured as highest elevated point on antiformal fold. (Axis of Anticline)

Trough - Lowest point of fold. (Axis of Syncline Fig. 7.2)

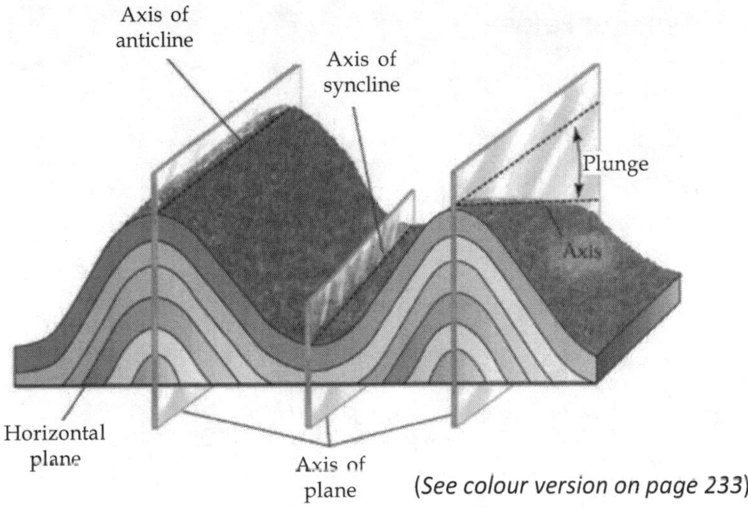

(*See colour version on page 233*)

Fig. 7.2

- Inter limb angle – The angle between two limbs is known as Inter limb angle.
- Axial plane surface – It is the surface that connects the hinge line of folded beds. It may be Vertical or inclined even horizontal which describe the type of fold geometrically.
- Anticline/Antiform – It is an upward pointing arch shaped fold in which beds are dipping in opposite directions. In an anticline, oldest rock is in the core and youngst at the peripheral part of the fold.
- Syncline/Synform – It is a basin shaped fold in which beds are dipping towards each other. In syncline older rocks are present at the periphery while youngst rock is at the center.
- Plunge – Normally, the fold axis is perpendicular to the horizontal plane. Plunge of fold is the angle of inclination of the axis or hinge line measured on a vertical plane.

7.4 Types of Folds

A. Basically, folds are of two types

a. Anticlinal Fold: When a fold exhibits convexity at the top, it is called an Anticlinal Folds.

b. Synclinal Fold: When a fold exhibits the concave side up then it is termed as Synclinal Fold. (Fig. 7.3).

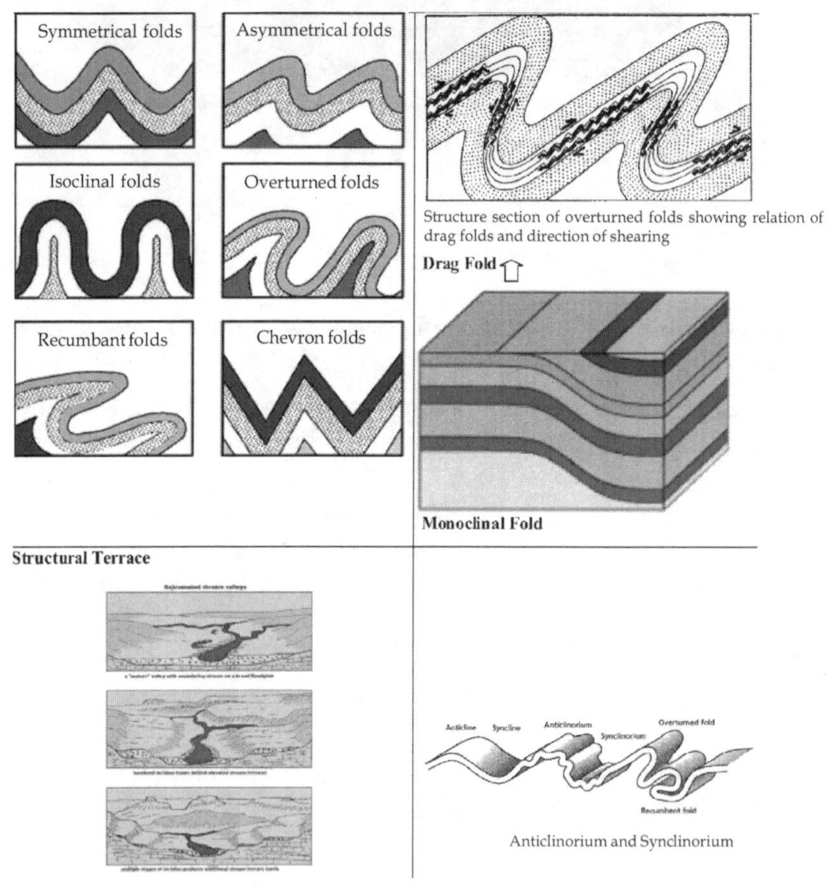

Fig.7.3 Types of folds

B. Based on nature of axial plane symmetry

1. **Symmetrical Fold**: When the two limbs of fold are inclined at the same angle on opposite direction, it is called symmetrical fold.

   In symmetrical folds, the axial plane is straight.

2. **Asymmetrical Fold**: When the two limbs of the fold are inclined at different angles and lengths it is called asymmetrical fold.

3. **Recumbent fold** : When axial plane is nearly horizontal. The fold is called Recumbent fold

4. **Overturned fold** : When the axial plane of a fold is inclined and both limbs are dipping in the same direction but showing different angles, it is called overturned fold.

5. **Isoclinal fold** : When both the limbs of fold are dipping at the same angle and in same direction that is they are parallel to each other the fold is termed as Isoclinal fold.

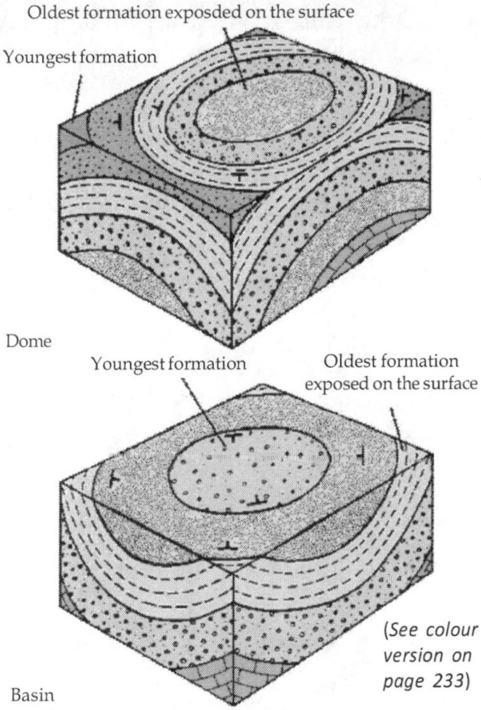

(See colour version on page 233)

Fig. 7.4 Dome and Basin

6. **Monoclinal fold** : Monocline means one limb of fold is inclined. When a horizontal bed becomes slightly inclined in part of fold show slight inclination and again becomes horizontal, then it is called monoclonal fold. It forms step like structure.

7. **Structural terrace** : It is local steepening in a uniform dipping strata and the flexure has a very gentle dip. Then it is called structural terrace.

8. **Fan fold** : When both the limbs are overturned, and give rise to fan like structure it is called Fan fold. In fan fold, crest and troughs are rounded.

9. **Chevron fold** : When the crest and trough of a fold are "angular" it is called Chevron fold (zig zag fold).

10. **Plunging fold** : When the hinge of fold is not horizontal and shows inclination, then it is called plunging fold. When the plunging exhibit inclination in two opposite direction then it is called doubly plunging fold. (Fig. 7.2)

11. **Drag fold** : When an incompetent (weak) bed lies between two competent beds and during tectonic activity the incompetent beds show minor asymmetrical folds, it is called drag fold.
12. **Basin** : Basin is a synclinal depression in which beds dip inward from all the directions. Beds commonly exhibit circular or elliptical outcrops on the surface (Fig. 7.4).
13. **Dome** : Dome is an anticline elevation in which beds dip outward from the centre. Dome commonly exhibit circular or elliptical outcrop.

7.5 Joints

Joints are the planar cracks and fractures with no displacement. They are formed by compressional and tensional forces on the earth's crust.

Joints are present in all types of rocks. In igneous rocks, joints are formed by the cooling and contraction of magma, therefore known as primary joints. Based on origins, its attitudes, joints are classified as follows:

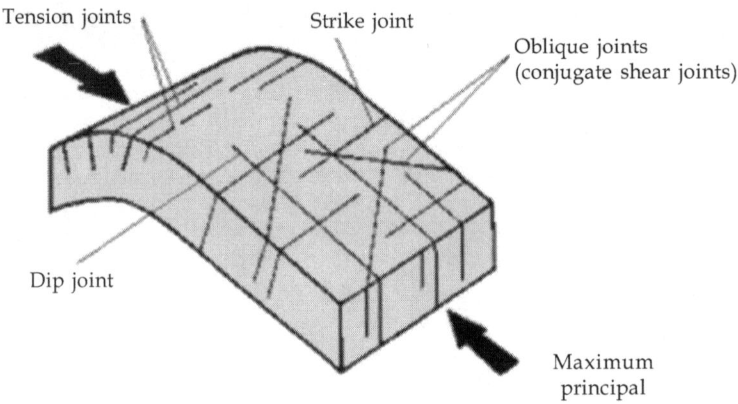

Fig. 7.5 Types of Joints

1. **Strike Joint** - If a joint plane is parallel to strike direction of sedimentary rocks or parallel to schistosity or gneissosity, it is called as strike joint.
2. **Dip Joint** - Joints parallel to dip of the planar structure of the rock.
3. **Oblique Joint** - If a joint is not parallel to dip or strike and placed oblique, it is called oblique joint.
4. **Columnar Joint** - This joint is mostly seen in Basalt. It is produced by cooling and contraction of magma and characterized by long hexagonal vertical columns.

5. **Bedding Joint** - When a joint is parallel to the bedding plane it is known as bedding joint.
6. **Conjugate Joint** - When two sets of joints are nearly parallel to ground surface. For example Granitic rocks.
7. **Sheet Joint** - Sheet joints are produced by physical or chemical processes of weathering. It develops exfoliation along center to several meter thick layers of rock (Fig. 7.6).

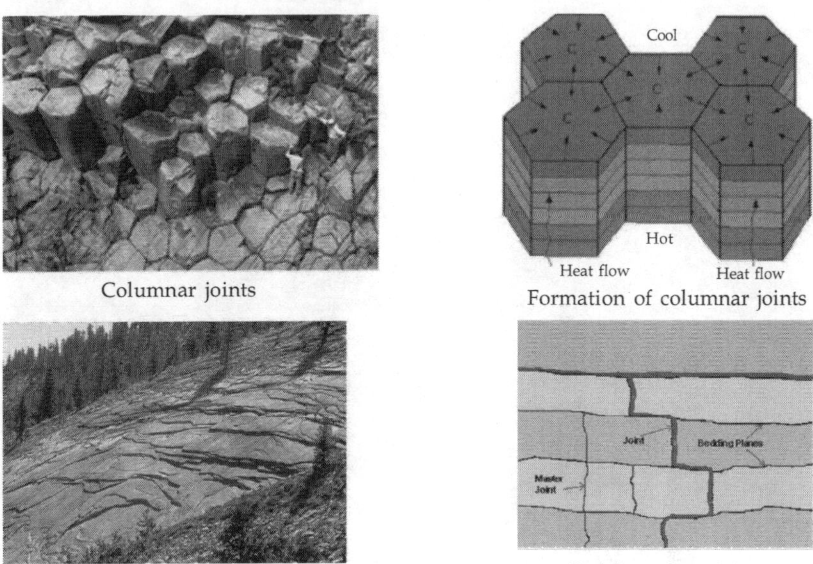

Columnar joints

Formation of columnar joints

Sheet joints

Bedding plane joints

Fig. 7.6 Types of Joints

(*See colour version on page 234*)

7.6 Unconformity

7.6.1 Definition

The term "unconformity" indicates a break in deposition of sedimentary layer. The unconformity surface represents a period of non-deposition, erosion or tectonic activity.

7.6.2 Types of unconformity

In nature, unconformities are of four types :

1. **Angular Unconformity** - When the two groups of rocks in contact are not parallel or showing some angle to each other, it is termed as "Angular Unconformity". Thus, the older beds show dipping while younger set of strata are horizontal or are conformable in disposition.
2. **Disconformity** - When the older bed and younger bed are horizontal but shows break in deposition, it is called "Disconformity".

3. **Local Unconformity** – It is the same as Disconformity but the areal extent is very less.
4. **Nonconformity** – When younger strata rest on eroded surface of igneous and or metamorphic rock, this is termed as "Nonconformity" (Fig. 7.7).

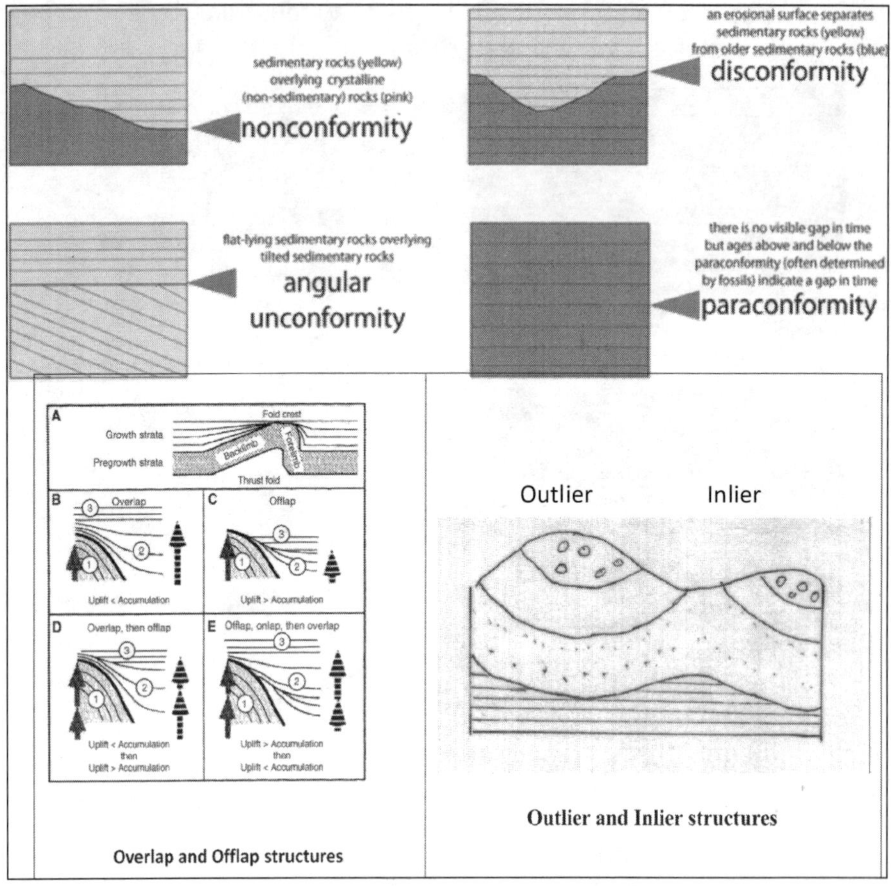

Fig. Types of Unconformity

7.6.3 Terms used in unconformity

1. **Unconformity plane** – The surface along which two sets of strata are joined on an unconformity surface is known as unconformity plane.
2. **Over-lap and Off-lap** – The structure is said to be "Over-lap" when the younger strata is seen extending beyond the underlying rock. When the younger set of strata is seen not extending on to the older set of strata, the structure is said to be "Off-lap".

3. **Outlier** - When the younger bed is surrounded by older beds, the structure is known as "Outlier".

4. **Inlier** - When the older bed is surrounded by younger ones the structure is known as "Inlier". The outliers and inliers are present in many areas in India. Example Jharia Coal Field, Jharkhand State.

7.6.4 Criteria to detect unconformity

There are certain criteria by which unconformities can be identified in field.

- Presence of conglomerate bed containing fragments of older rock.
- Presence of erosional, irregular and undulating contact between two sets of strata.
- Presence of different flora or fauna.
- Truncation of igneous intrusive suite of rocks by the younger set of strata or rocks.
- Differences in deformational structure or metamorphism in the two sets of strata placed adjacentely.

7.7 Faults

Faults are planar or curviplanar fractures in the rocks/strata forming the earth crust along which displacement or relative movement has taken place is known as "Fault".

7.7.1 Fault terminologies (Fig. 7.8)

1. **Fault plane** - The surface along which movement takes place is known as fault plane.
2. **Hanging wall** - It is the portion of fault that is above the inclined fault plane.
3. **Foot wall** - It is the part of fault that is lying below the inclined to fault plane or/to hanging wall.
4. **Translational movement** -The movement of faulted set of strata along and across the dip without any rotation.
5. **Rotational movement** - The movement in which one block has rotated other.
6. **Throw of fault** - Vertical displacement between the up thrown and down thrown blocks of a fault.

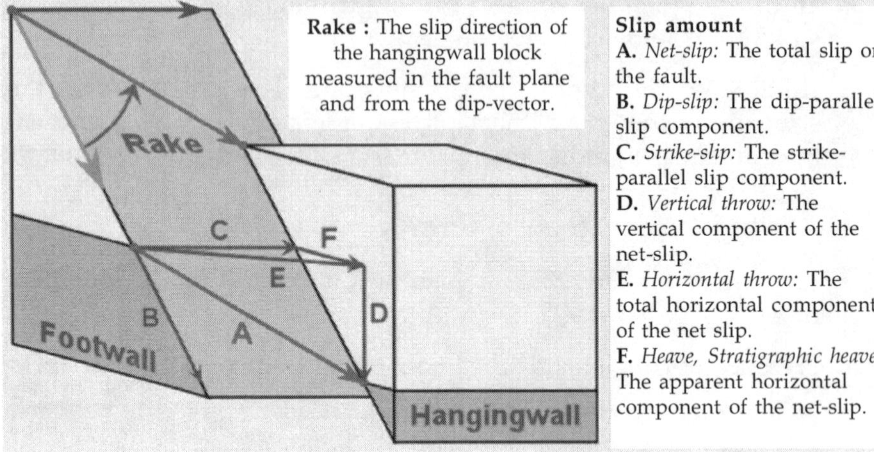

Fig. 7.8 Fault terminologies

7. **Hade** – The angle of inclination of fault plane is known as Hade.
8. **Heave** – The apparent horizontal component of the net slip of the fault is known as Heave.

7.7.2 Types of faults

A. The classification of fault is based on the movement along the fault plane and its relationship to the dip and strike of fault plane.

1. **Strike slip fault** – When the direction of displacement is parallel to strike of fault. It is strike slip fault.
2. **Dip slip fault** – The direction of displacement is parallel to dip of fault (upward or downward). It is dip slip fault.
3. **Oblique fault** – A fault is said to be oblique when the direction of displacement is not parallel to strike or dip of fault plane.

B. Based on displacement

1. **Normal fault** – A fault is said to be normal when the displacement of hanging wall is in the direction of the fault plane inclination.
2. **Reverse fault** – It is opposite to normal fault when hanging wall moves upward relatively.
3. **Step fault** – When a series of closely spaced faults showing down throw movement in the same direction, it is known as "step fault".

4. **Pivotal/Scissor fault** - When the displacement takes place along fault plane from a pivot, rotated and thrown towards other side of the pivot.

5. **Horst and Graben** - Horst structure is when elongated blocks of rock elevated or showing upward movement along parallel fault planes, while Graben structure shows downward movement and forming depressions. It creates a long valley and also known as rift valley (Fig. 7.9).

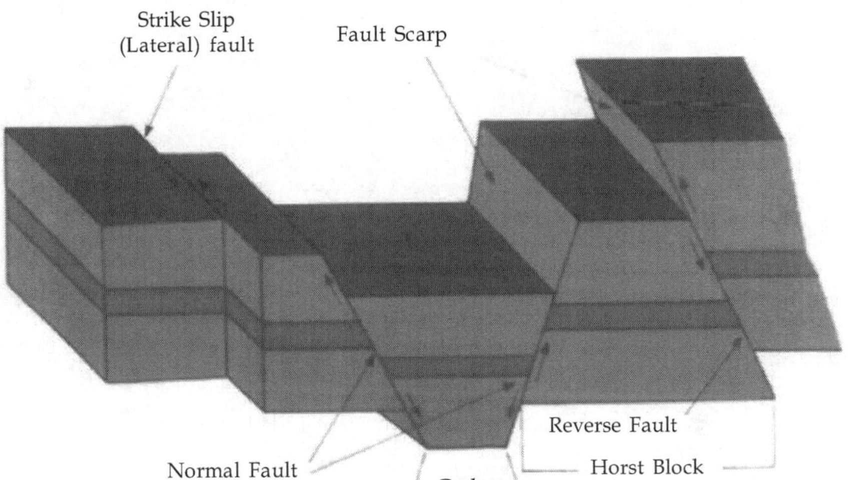

Fig. 7.9 Types of Fault

7.6.3 Recognition of faults

- Repetition of bed
- Sudden disappearance of bed/strata.
- Presence of smooth surface or slickensides represent fault plane
- Presence of fault breccia
- Presence of spring (Geomorphic criteria) specially hot springs.

Chapter 8

Economic Geology, Ore Mineralogy, Remote Sensing and Photogeology

8.1 Economic Geology and Ore Mineralogy

8.1.1 Economic geology

Economic Geology deals with the comprehensive study of mineral deposits including their processes of formation and uses. By applying the principles of geology, a mining engineer is able to exploit such deposits profitably for the use of Man.

8.1.2 Ore mineral

An ore is a term applied usually to that part of a metalliferous mineral deposit that can be utilized profitably as a source of one-or-more metals, this term now is widely used for the usable part of many non-metallic (non-metalliferous mineral) deposits, like barite, fluorite, sulphur, etc.

8.1.3 Gangue mineral

An ore is usually a mixture of one or more usable minerals (ore minerals) and one or more unusable minerals (gangue minerals). The usual gangue minerals are quartz and other forms of silica, calcite, dolomite, siderite, barite, feldspars, garnet, chlorite, kaolin, fluorite, apatite, pyrite, pyrrhotite, arsenopyrite, and sometimes the gangue material is the country rock or host rock itself in which the ore minerals occur. Sometimes, the gangue minerals may also be utilized as by-products of the ore deposit. The nature and proportion of the gangue influence the economic value of an ore to a great extent.

8.1.4 Tenor of ore

The metal content of an ore is called the tenor of the ore. It is generally expressed in percentage of the metal (in ounce or penny weight per ton in case of precious metals). There are about 200 common economic minerals, and of these some are necessary for essential industries, others are used in non-essential industries. Some minerals are processed for materials of luxury, whereas others are vital for purposes of defence. Many of the economic minerals are used in more than one industry, and hence there is an overlapping in their respective nature of uses. The economic minerals, are broadly classified into metallic and non-metallic, the latter are also called industrial minerals and include some almost mono-mineralic rocks like limestone, dolomite, quartzite, rock-san and gypsum and building stones and road metals,. The 'fossil fuels'-coal and petroleum which are not strictly minerals, are included in fuel minerals.

8.1.5 Metallic minerals

The metalliferous (metallic) minerals which are mainly used in the extraction of metals are classified into (i) Ores of iron and alloy metals (Fe, Mn, Cr, Ni, Co, Mo, W, V and Ti); (ii) Ores of non-ferrous metals (Al, Cu, Pb, Zn, Sn); (iii) Ores of precious metals (Au, Ag, Pt) and (iv) Ores of other industrial metals and submetals (Be, Mg, Nb, Ta, Th, U, Zr, Cd, As, Bi, Sb, Hg).

8.1.6 Non metallic minerals

The industrial (non-metallic) minerals are classified according to the industries (other than, or in addition to, the extraction of metal) in which they are used.

8.1.7 Physical and optical properties of Ore minerals

Metallic minerals are opaque in nature and can not be studied under normal polarizing microscope like rock and minerals. Polished sections of opaque minerals are studied in reflected light by ore microscope.

1. Name of Mineral – Pyrolusite

A. Physical Properties

1. Colour	Iron grey or dark steel grey
2. Form	Massive, pseudomorphous, reniform
3. Cleavage	Absent
4. Fracture	Uneven or brittle
5. Lustre	Sub-metallic
6. Hardness	2 to 2.5

7.	Specific Gravity	4.8
8.	Crystal System	Tetragonal
9.	Diagnostic Properties	Iron grey colour, brittle substance with 4.8 sp gravity.
10.	Mode of Occurrence	Occurs as syngenetic mineral in Kodurite rock or as replacement deposits in Gondite.
11.	Distribution	Mn deposits found in M.P., Maharastra, A.P. Karnataka, Gujarat, Rajasthan and Goa. Mn is associated with Gondite deposits in Sausar Series, Bharweli mines, Balaghat, Chhindwara, Jabalpur, Jhabua district in MP and Bhandara and Nagpur district of Maharastra. Mn associated with Kodurite rocks in Visakhapatnam, Srikakululam, Adilabad and Hyderabad district in AP and Koraput and Bolangir district in Orissa.
12.	Economic Use	• It is used as ore mineral of Mn. • When mixed with Fe, it forms hard steel with extra resistance. • Used in manufacturing of Dry Batteries. • Used in Glass Industries for decoloring agent.
13.	Name of Mineral	Pyrolusite MnO_2

Ore microscopy: Polished section

1.	Crystal System	Tetragonal.
2.	Color	Creamy white.
3	Bireflectance/ Pleochroism	Yellowish white to grayish white.
4	Anisotropy	Very strong, Yellowish brownish blue.
5.	Internal Reflectance	Not present.
6.	Polishing Hardness	Very variable, depending upon grain size and orientation.
7.	Name of mineral	Pyrolusite.

2. Psilomelene

Physical Properties

	Property	Description
1.	Colour	Iron black
2.	Form/habit	Amorphous, massive, botryoidal (Bunch of grapes) or reniform (Kidney shaped)
3.	Cleavage	Absent
4.	Fracture	Sub-conchoidal to uneven
5.	Lustre	Sub-metallic, dull
6.	Hardness	5-6
7.	Specific Gravity	3.7-4.7
8.	Crystal System	Tetragonal
9.	Diagnostic Properties	Form, Habit
10.	Mode of Occurrence	It commonly occurs as botryoidal masses of very fine acicular crystals in concentric layers
11.	Economic Use	Ore of Manganese
12.	Name of Ore mineral	Psilomelane.

Ore microscopy: Polished section

	Property	Description
1.	Crystal System	Massive Hard Mn Oxide.
2.	Colour	Bluish Grey.
3.	Bireflectance/ Pleochroism	Strong with bluish grey.
4.	Anisotropy	Strong white to grey.
5.	Internal Reflectance	Occasional brown.
6.	Polishing Hardness	Variable, depending upon grain size and orientation.
7.	Name of mineral	Psilomelane.

3. Magnetite (Fe_3O_4)

A. Physical Properties

1.	Colour	Iron Black
2.	Form/habit	Crystalline or massive
3.	Cleavage	Imperfect
4.	Fracture	Sub conchoidal
5.	Lustre	Metallic
6.	Hardness	5.5-6.5
7.	Specific Gravity	5.18
8.	Crystal System	Isometric Octahedral
9.	Diagnostic Properties	Iron black color, high sp gravity crystalline form
10.	Mode of Occurrence	Magnetite is associated with BHQ and BHJ of Iron Ore Series, in Jharkhand and Odisha. Dharwar Group in Karnataka, Archaean age. It occurs as euhedral, subhedral and anhedral polycrystalline aggregates associated with pyrhotite, pyrite, pentlendite, chalcopyrite, bornite, sphelerite, galena. etc.
11.	Distribution in India	Important Magnetite deposits in India are; i) Kudremukh, Chickmanglur, Karnataka, ii) Salem, Tamil Nadu iii) Bailadila and Dallirajhara of Chhattisgarh iv) Noamundi and Gua of Jharkhand v) Kyonjhar, Mayurbhanj and Sundergarh districts, Odisha vi) Ratnagiri Bhandara and Chandrapur districts of Maharastra
12.	Economic Use	It is an ore of Iron. Used in making Pig Iron, Steel
13.	Name of mineral	Magnetite.

Ore microscopy: Polished section

1.	Crystal System	Cubic
2.	Colour	Grey with brownish tinct
3.	Bireflectance/ Pleochroism	Not present
4.	Anisotropy	Isotropic
5.	Internal Reflectance	Not present.
6.	Polishing Hardness	> pyrrhotite, < Ilmenite, Haematite, Pyrite.
7.	Diagnostic Properties	Colour, Isotropic.
8.	Name of mineral	Magnetite.

4. Name of Mineral – Hematite (Fe_2O_3)

1.	Color	Steel grey to iron black
2.	Form/habit	Massive sometimes granular or micaceous with flacks
3.	Streak	Cherry red
4.	Cleavage	Absent.
5.	Fracture	Uneven to sub conchoidal
6.	Lustre	Metallic, dull sometime shining
7.	Hardness	5.5 to 6.5
8.	Specific Gravity	4.9 to 5.3
9.	Crystal system	Hexagonal
10.	Diagnostic Properties	Cherry red streak.
11.	Mode of Occurrence	Hemetite is associated with BHQ and BHJ of Iron Ore Series, Dharwar group, Archaean age. It occurs as euhedral, subhedral and anhedral polycrystalline aggregates associated with pyrhotite, pyrite, pentlendite, chalcopyrite, bornite sphelerite galena. etc.

12.	Distribution in India	Important deposits in India are; vii) Kudremukh, Chickmanglur, Karnataka, viii) Salem Tamil Nadu ix) Bailadila & Dallirajhara of Chhattisgarh x) Noamundi and Gua of Jharkhand xi) Kyonjhar, Mayurbhanj and Sundergarh or Odisha xii) Ratnagiri Bhandara and Chandrapur District of Maharastra
13.	Economic Use	It is an ore of Iron. Used in making Pig Iron, Steel
14.	Name of Mineral	Hematite.

Ore microscopy: Polished section

1.	Crystal System	Hexagonal
2.	Colour	Grey white with bluish tint
3.	Bireflectance/ Pleochroism	Weak
4	Anisotropy	Distinct grey
5.	Internal Reflectance	Deep red common
6.	Polishing Hardness	> Magnetite, < Pyrite
7.	Diagnostic Properties	Colour, Internal reflectance
8.	Name of mineral	Hematite

5. Name of Mineral – Ilmenite ($FeTiO_2$)

A. Physical Properties

1.	Colour	Iron black
2.	Form/habit	Crystalline or massive
3.	Cleavage	Parting
4.	Fracture	Conchoidal
5.	Lustre	Sub-metallic

6.	Hardness	5 to 6
7.	Specific Gravity	4.5 to 5.0
8.	Crystal System	Trigonal
9.	Diagnostic	Iron black, crystalline, hardness 5-6
10.	Mode of Occurrence	Ilmenite is commonly found in beach sands on coastal tracts Mostly occurs as titaniferous magnetite in Igneous rock
11.	Distribution	Rich deposits of Ilmenite is found in coastal tracts of Ratnagiri in Maharastra, Kollam Kerala, Kanya Kumari Thanjavur district of Tamil Nadu, Cuttack and Ganjam distt, Odisha
12.	Economic Use	It is an ore of Titanium
13.	Name of mineral	Ilmenite.

Ore microscopy: Polished section

1.	Crystal System	Tetragonal.
2.	Colour	Brownish violet and pink tint.
3.	Bireflectance/ Pleochroism	Distinct pinkish.
4.	Anisotropy	Strong, greenish grey to brownish grey.
5.	Internal Reflectance	Rare dark brown.
6.	Polishing Hardness	> Magnetite < Haematite.
7.	Diagnostic Properties	Colour and Bireflectance.
8.	Name of mineral	Ilmenite.

6. Galena (PbS)

Physical Properties

1.	Colour	Lead grey
2.	Form/habit	Crystalline, cubic, also occur as massive

3.	Streak	Lead grey
4.	Lustre	Metallic, splendent
5.	Cleavage	3 sets cubic, perfect one set (100)
6.	Fracture	Sub-conchoidal
7.	Hardness	2.5
8.	Specific Gravity	7.4 to 7.6 (High)
9.	Crystal System	Cubic
10.	Diagnostic Properties	Form, cleavage, Hardness and high Sp. Gravity.
11.	Mode of Occurrence	It occurs as vein in limestone, sandstone, occasionally in metamorphic or volcanic rocks
12.	Distribution	Galena is associated with Zn deposits in Rajasthan. Important are Zawar Group of Mines, Udaipur. Galena is associated with dolomitic limestone at Mochia Mogra. Ambamata Polymetallic deposits at Gujarat.
13.	Economic Use	Ores of Pb, Used in manufacturing of Storage Batteries, Used as alloy with other minerals, In making Pigments
14.	Name of mineral	Galena.

Ore microscopy: Polished section

1.	Crystal System	Cubic
2.	Colour	White with pink tinct
3.	Bireflectance/ Pleochroism	Not present
4.	Anisotropy	Isotropic weak
5.	Internal Reflectance	Not present
6.	Polishing Hardness	Chalcopyrite, < Tetrahedrite
7.	Diagnostic Properties	Form- Cubic, Isotropic
8.	Name of mineral	Galena

7. Realgar (AsS)

1.	Colour	Red or orange
2.	Form/habit	Usually massive or granular, rarely occurs as prismatic crystal
3.	Streak	Red or orange
4.	Lustre	Metallic, splendent
5.	Cleavage	Perfect 3 sets cubic
6.	Fracture	Conchoidal uneven
7.	Hardness	1.5 - 2.0
8.	Specific Gravity	3.56
9.	Crystal System	Monoclinic
10.	Diagnostic Properties	Orange colour, granular form with low hardness 1.5 to 2
11.	Mode of Occurrence	Occurs as irregular plate like masses with orpiment, also associated with Stibnite, Arsenopyrite, Pyrite etc.
12.	Distribution	It occurs in Bihar, Haryana, Jammu Kashmir, Karnataka
13.	Economic Use	Ore of Arsenic. Used in glass industry, insecticides, used in making alloys
14.	Name of mineral	Realgar

Ore microscopy: Polished section

1.	Crystal System	Monoclinic
2.	Colour	Dull grey
3.	Bireflectance/ Pleochroism	Weak but distinct, grey with reddish to bluish tinct
4.	Anisotropy	Strong
5.	Internal Reflectance	Abundant and Intense, Yellowish red
6.	Polishing Hardness	< Orpiment,
7.	Diagnostic Properties	Colour, Birefrectance
8.	Name of mineral	Realgar

8. Orpiment (As_2S_3)

1.	Colour	Fine lemon, yellow
2.	Form/habit	Usually massive or granular, rarely crystalline
3.	Streak	Yellow
4.	Lustre	Resinous
5.	Cleavage	Perfect one set in crystal (Prismatic 010)
6.	Fracture	Conchoidal
7.	Hardness	1.5 to 2
8.	Specific Gravity	3.4 to 3.5 light
9.	Crystal System	Monoclinic
10.	Diagnostic Properties	Lemon yellow, massive form with low hardness 1.5-2
11.	Mode of Occurrence	It occurs as tabular interlocking anhedral masses, and as needles or lath like crystals, often formed on Realgar. Association with Stibnite, Arsenopyrite, Pyrite etc is common.
12.	Economic Use	Ore of Arsenic
13.	Name of mineral	Orpiment

Ore microscopy: Polished section

1.	Crystal System	Monoclinic
2.	Colour	Grey
3.	Bireflectance/ Pleochroism	Strong
4.	Anisotropy	Strong
5.	Internal Reflectance	Abundant and strong
6.	Polishing Hardness	> Realgar
7.	Diagnostic Properties	Colour, Bireflectance, Anisotropy
8.	Name of mineral	Orpiment

9. Chalcopyrite ($CuFeS_2$)

1.	Colour	Brass yellow often with iridescent tarnish
2.	Form/habit	Twinned crystals
3.	Streak	Greenish black, shining
4.	Lustre	Resinous
5.	Cleavage	Poor (011) and (111)
6.	Fracture	Conchoidal to uneven
7.	Hardness	3.5-4.0
8.	Specific Gravity	4.1-4.3
9.	Crystal System	Tetragonal
10.	Diagnostic Properties	Colour, Streak, Hardness, Sp Gravity
11.	Mode of occurrence	It mainly occurs as porphyry copper or as hydrothermal veins within various types of rocks and tectonic environments. It is also associated with Fe pyrite in sandstone.
12.	Economic Uses	• Important ore minerals of Copper • Due to high electrical conductance, Cu is used in Electrical Industries • Used in the production of Alloy metal like bronze and brass • Its oxides, sulphates and sulphides are used as a fungicide and coloring of glass
13.	Distribution	Important occurrence of Copper Deposits are as follows: • Khetri Copper Complex, Distt. Jhunjhunu, Rajasthan • Malajkhand Copper Deposit, Balaghat, M.P. • Mosabani and Rakha Mines of Singhbhum, Jharkhand
14.	Name of mineral	Chalcopyrite.

Ore microscopy: Polished section

1.	Crystal System	Tetragonal
2.	Colour	Yellow to brassy yellow
3.	Bireflectance/ Pleochroism	Weak
4.	Anisotropy	Weak but distinct
5.	Internal Reflectance	Not present
6.	Polishing Hardness	Galena < Sphalerite
7.	Diagnostic Properties	Colour, polishing hardness
8.	Name of mineral	Chalcopyrite

10. Bornite (Cu_5FeS_4)

1.	Colour	Copper red to golden brown
2.	Form/habit	Usually massive
3.	Streak	Pale greyish black
4.	Cleavage	Perfect one set (111)
5.	Fracture	Conchoidal or uneven
6.	Hardness	3.0
7.	Specific Gravity	4.9-5.4 (High)
8.	Crystal System	Tetragonal
9.	Diagnostic Properties	Colour, Hardness and Specific Gravity
10.	Mode of Occurrence	Occurs as polycrystalline aggregates with Chalcopyrite
11.	Distribution	Same as Chalcopyrite
12.	Economic Use	Same as Chalcopyrite
13.	Name of mineral	Bornite

Ore microscopy: Polished section

1.	Crystal System	Tetragonal
2.	Colour	Pink, brown to orange
3.	Bireflectance/ Pleochroism	Visible in grain boundaries
5.	Anisotropy	Very weak
6.	Internal Reflectance	Not present
7.	Polishing Hardness	> Galena, < Chalcopyrite
8.	Diagnostic Properties	Crystal system, Polishing Hardness
9.	Name of minral	Bornite

11.Covellite (CuS)

1.	Colour	Indigo blue
2.	Form/habit	Platy hexagonal crystals usually massive
3.	Streak	Lead grey to black
4.	Lustre	Sub metallic to Resinous
5.	Cleavage	Basal perfect
6.	Fracture	Conchoidal
7.	Hardness	1.5 to 2
8.	Specific Gravity	4.6
9.	Crystal System	Hexagonal
10.	Diagnostic Properties	Color, hardness and specific gravity
11.	Mode of Occurrence	Occurs as subhedral to anhedral masses, Brilliant colour and strong pleochroism and anisotropism are diagnostic characters. Associated with copper and iron sulphides pyrite, chalcopyrite and bornite etc.
12.	Distribution	Same as Chalcopyrite and Bornite
13.	Economic Use	Same as Chalcopyrite and Bornite
14.	Name of mineral	Covellite.

Ore microscopy: Polished section

1.	Crystal System	Hexagonal
2.	Colour	Creamy white
3.	Bireflectance/ Pleochroism	Purple to violet red
4.	Anisotropy	Red orange to brownish
5.	Internal Reflectance	Not present
6.	Polishing Hardness	< Chalcopyrite
7.	Diagnostic Properties	Color, Bireflectance, Polishing Hardness.
8.	Name of mineral	Covellite

8.1.8 Classification of mineral deposits according to their origin

Mineral deposits may be mainly classified into the following two broad groups according to the two environments in which they are formed.

1. **Endogenetic Mineral Deposits** are formed by the processes originating or operating within the lithosphere or the crust. Some of these deposits result from the progressive crystallisation of a magma, its interaction with the country rocks or from its exuding emanations consisting of gases, vapours and liquids; such deposits may be regarded as hypogene or primary deposits.

2. **Exogenetic Mineral Deposits** are formed at or near the earth's surface by processes, involving integrated action of hydrophere and atmosphere on the lithosphere. They comprise all the secondary deposits such as residual, deposits formed by circulating meteoric waters, oxidized and secondarily enriched deposits, and sedimentary deposits etc.

A) Endogenetic Deposits

These are classified into the following six main types:

1. **Magmatic deposits**: These result from simple crystallisation or magmatic segregation of minerals in the earliest stage (orthomagmatic stage) of crystallisation of slowly-cooling magma (early magmatic deposits), or from concentration and subsequent injection of the residual ore-forming fraction of that magma after

the pyrogenic rock-forming minerals had crystallised (late magmatic deposits).

2. **Pegmatitic:** Pneumatolytic Deposits are derived from the crystallisation of the pegmatitic differentiate of a magma which is the less viscous/more mobile volatile-rich residual fraction in which rare elements also become highly concentrated. Such pegmatitic liquids may be squeezed out to fill in the cracks and fissures in the parent igneous body or in the adjoining country rocks, and form pegmatite veins. The remaining volatile fraction containing abundant chemically active elements may deposit mineral substances in the pre-existing openings of the rocks by filling or may form metasomatic replacement bodies in the rocks, and give rise to pneumatolytic deposits.

3. **Hydrothermal deposits:** They are formed by hydrothermal solutions-hot ascending hydrous magmatic emanations charged with mineral matter, which are capable of transporting mineral matter to great distances from the parent igneous mass. They are mostly epigenetic deposits, i.e. the ore bodies are introduced subsequently into the pre-existing rock. They occur as cavity fillings in pre-existing rock openings, cracks, fissures and bedding, foliation and fault planes, or they may be formed by replacement of the country rock encountered along the path of the hydrothermal solutions.

   On the basis of temperature and pressure, the depth of their formation and the distance from the magmatic source, the hydrothermal deposits are subdivided into five types.

i) **Hypothermal deposits**: They are formed by hot hydro-thermal solutions at great depth characterised by high pressure and comparatively high temperature (3000-5000C). They occur in, or in close proximity of, the parent intrusive bodies of deep-seated origin.

ii) **Mesothermal deposits**: They typically form at intermediate depths (1200m to 3650m below surface) and temperatures (2000-3000C). In these deposits, fillings of cavities and fractures is more dominant than replacement, and they occur generally not very far off from the intrusive rocks.

iii) **Epithermal depostis**: These are formed at shallow depth (less than 900m below surface) under moderate pressure and low temperature (500-200°C). They are localized comparatively in younger rocks as cavity-filled deposits, fissure veins, etc., being characterized by crustification, comb structure and drusy cavities. The typical ore-minerals of epithermal deposits are cinnabar, stibnite, realgar, orpiment, etc.

iv) **Telethermal deposits** : These are also considered to have formed hydrothermal solutions that have migrated for such long distances from their parent igneous bodies that their genetic relationship with them is not properly established. These deposits are formed at shallow depth under low temperature, and are characterized by lesser intensity of wall-rock alteration effects than the other above-mentioned hydrothermal types of mineral deposits.

4. **Xenothermal deposits** : These are formed by high temperature ore-forming fluids expelled from huge igneous rock masses which have intruded into shallow depths. Such deposits are formed under conditions of rapid cooling and sudden losses of pressure of the ascending fluids, and are, therefore, characterized by mixed high-temperature and low-temperature mineral suites and are consequently not easily recognized by temperature pressure criteria.

5. **Fumarolic deposits** : These are sublimated and deposited at or near the surface at low temperature and pressure due to the sudden cooling of the vapours emanating from volcanoes or fumaroles. Sulphur is the typical example of this class of deposits.

6. **Deposits formed by metamorphic processes** : These are formed by processes of metamorphism, the chief agencies being heat, pressure and water. There include the following two types of deposits.

i) **Metamorphosed mineral deposits** : These are formed as a result of metamorphism of preexisting mineral deposits, due to which the ores acquire new textural characters like granular, schistose, banded and flow textures. Moreover, the original ore minerals may be dehydrated, and transformed into new mineral. Certain iron and manganese ore deposits have been formed by metamorphism of ferruginous and manganiferous sediments respectively.

ii) **Metamorphic mineral deposits** : These include some deposits of non-metallic minerals likegraphite, garnet, kyanite-sillimanite, asbestos, steatite, etc. which are formed by the recrystallisation and/ or recombination of certain mineral constituents of the rocks during regional or contact metamorphism.

B) Exogenetic Deposits

These may be classified into the following four main types:

1. **Deposits by meteoric circulating water** : These deposits are formed by ground water at shallow depths by the solution and concentration of material dispersed in the surrounding rocks or deposits. These

deposits, occur mainly in sedimentary rocks-limestones, sandstones and shales in which carbonaceous matter or other reducing agensts may be present.

2. **Oxidised and secondarily enriched ore deposits** : These are formed by processes of weathering, leaching and chemical changes arising from the action of descending meteoric waters on a pre-existing mineral deposits. In such deposits the primary minerals are oxidised and weathered at the surface resulting in the formation of 'gossan' (weathered zone, which may contain sometimes few stable residual ore minerals). Just below this, there is a zone of leaching (mostly devoid of ore minerals) followed by zone of oxidation which is just above the water table and contains ore minerals, such as native metals, oxides, carbonates and hydrous silicates. Below the water table is the zone of secondary or supergene enrichment, the most important zone, in which reaction between the primary minerals and the descending solutions of oxy-salts results in deposition of minerals richer in the metal contents than the primary ore-minerals. Below this zone lies the deposit in its primary condition. The common example are copper loades, containing chalcopyrite and iron pyrite.

3. **Residual deposits** : They are formed by the rock-decay and weathering near the surface under low or moderate pressure and at temperatures varying from 0-100°C. They represent the valuable minerals left behind or formed in more or less concentrated form-after the removal of the other undesirable constituents of the rock, and overlie the unweathered parent rock. Examples are deposits of gold, cassiterite, bauxite, iron ore, manganese ore, kyanite etc.

4. **Sedimentary deposits** : These mineral deposits are formed at or near the surface by the same processes that form the sedimentary rocks. They are all syngenetic, i.e., they are formed as a part of the strata and are contemporary with the rest of the rock. They may be grouped in the following three types:

i) **Detrital deposits** : These are sedimentary deposits formed by mechanical processes of erosion, transportation and deposition during which concentration of a valuable mineral may take place. These deposits include sediments and sedimentary rocks like clays, sands, gravels, sandstone, conglomerates, and also placer deposits like those of gold, cassiterite, wolframite, ilmenite, monazite, etc., in which the heavy and more resistant valuable minerals, present only in small amount in rocks are concentrated by the selective transportation due to gravity, running water (stream placers) or wind (eolian placers). Placer deposits usually form along hill slopes (eluvial deposits), along stream channels (stream-deposits) or along beaches (beach deposits).

ii) **Chemical-organic deposits** : These are deposits formed chemically due to reactions between solutions in bodies of surface water or by the accumulations of organic matter. Plants and animals play an important part in the concentration of certain elements in the surface water and their subsequent precipitation as mineral deposits. These deposits include beds of limestone, dolomite, limonite, haematite, siderite, phosphorite etc., and also the coal seams and the 'oil sands'.

iii) **Evaporation deposits** : These are residues (precipitates of soluble salts) formed by the evaporation of surface waters (of seas, bays and lakes) like deposits of rock-salt, gypsum and anhydrite etc. In this class may also be included the deposits made by spring waters, as well as the efflorescent or saline crust formed due to the evaporation of salt solutions brought to the surface from below by capillary action in dry season, as is the case with the 'reh' and the saltpetre (nitre) deposits of northern India.

8.1.9 Distribution of important metallic and non metallic minerals in India

(A) Distribution of Iron Ore in India

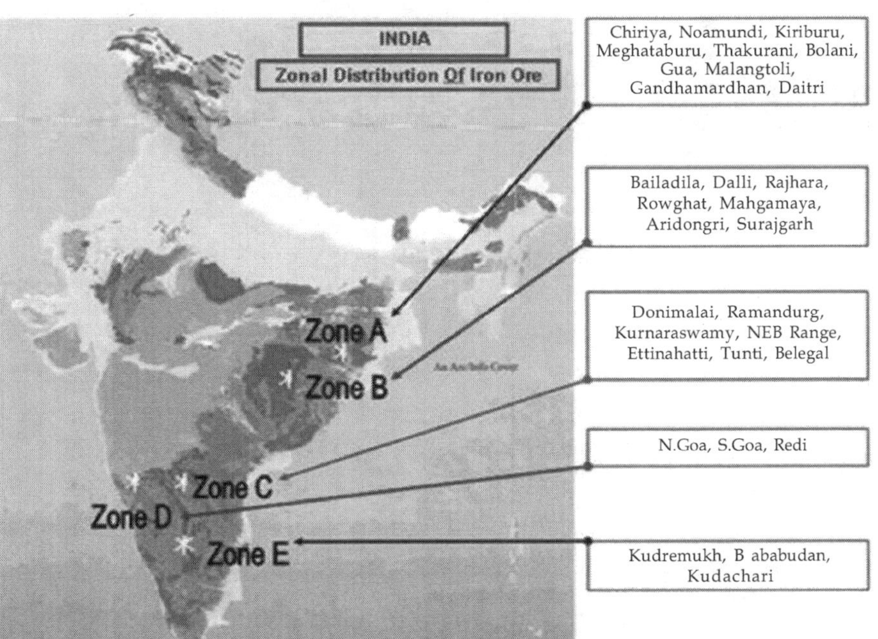

Distribution of Iron Ore in map of India

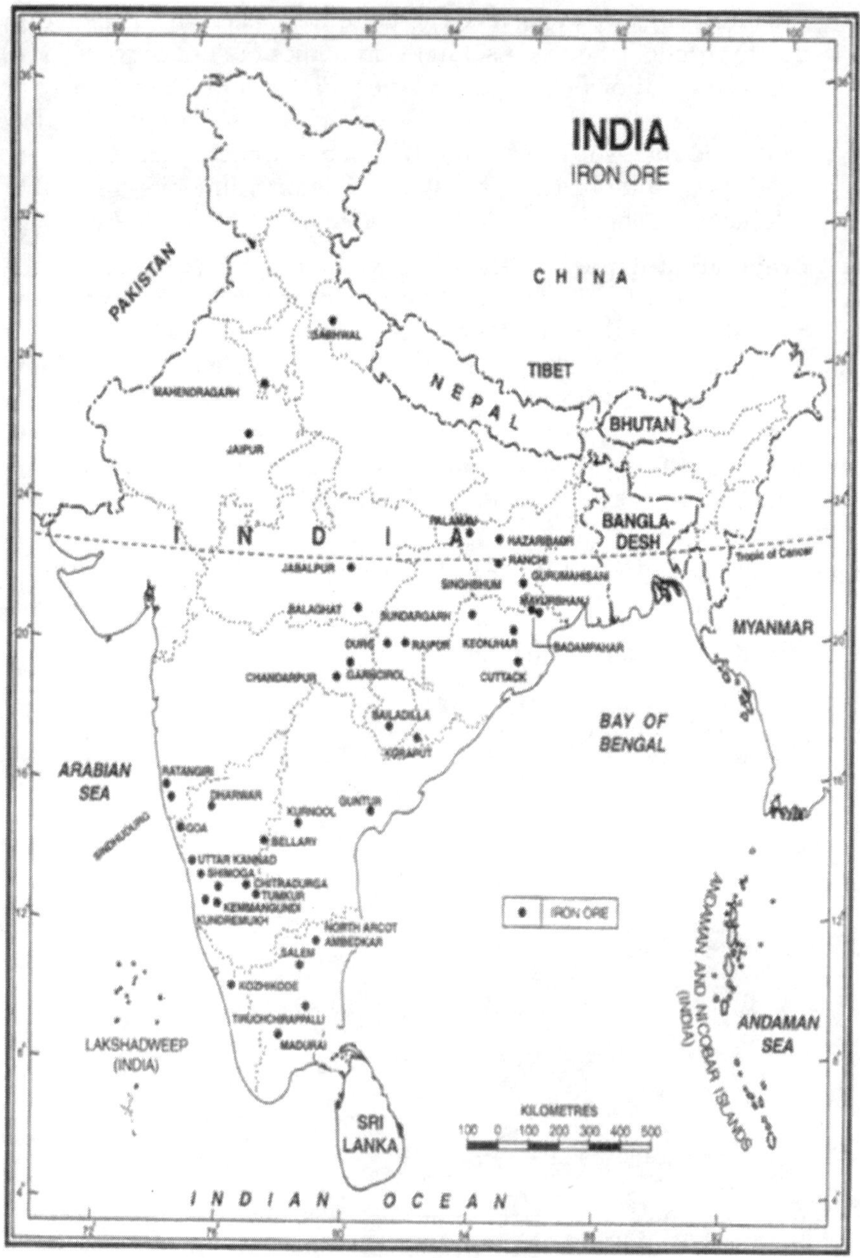

Distribution of Mn Deposits in Map of India

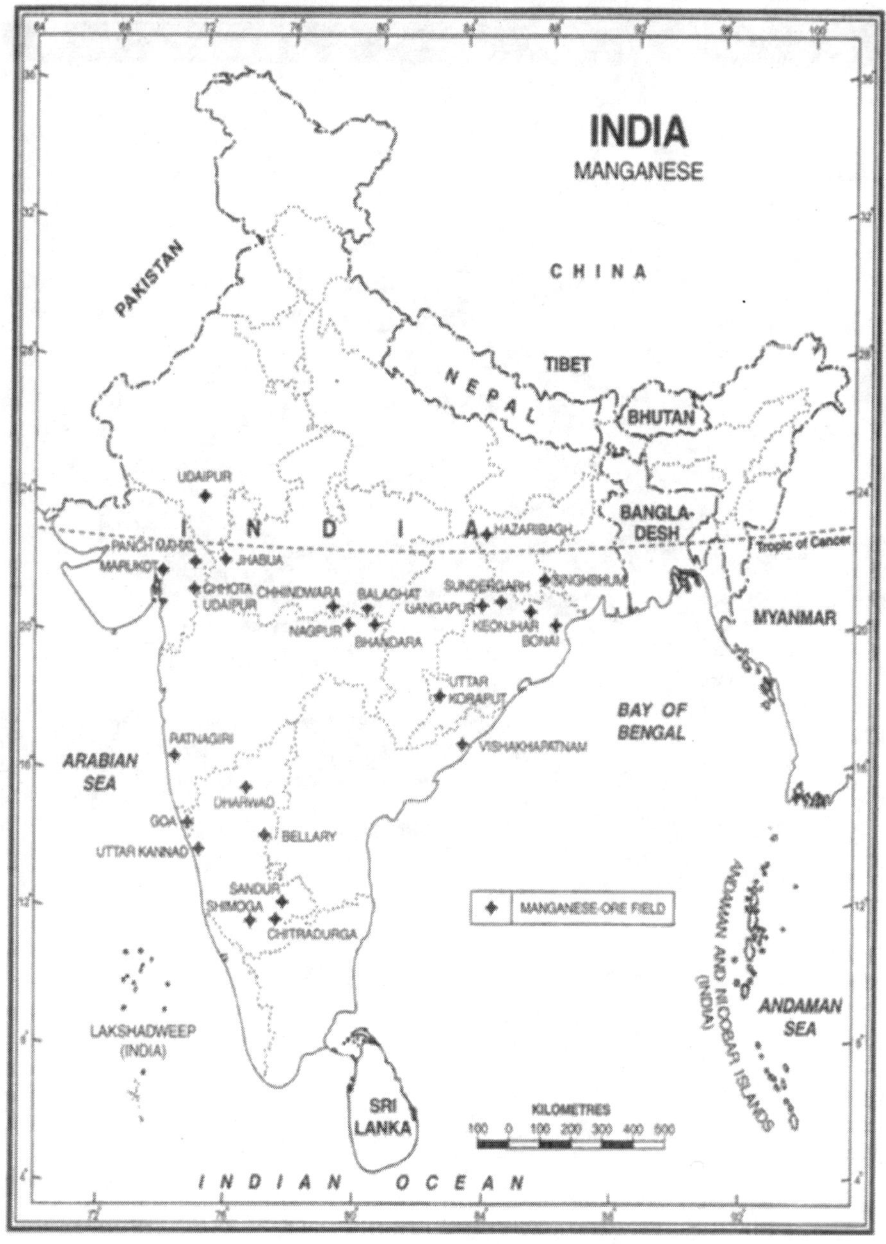

Distribution of Copper and Aluminium Ore in the map of India

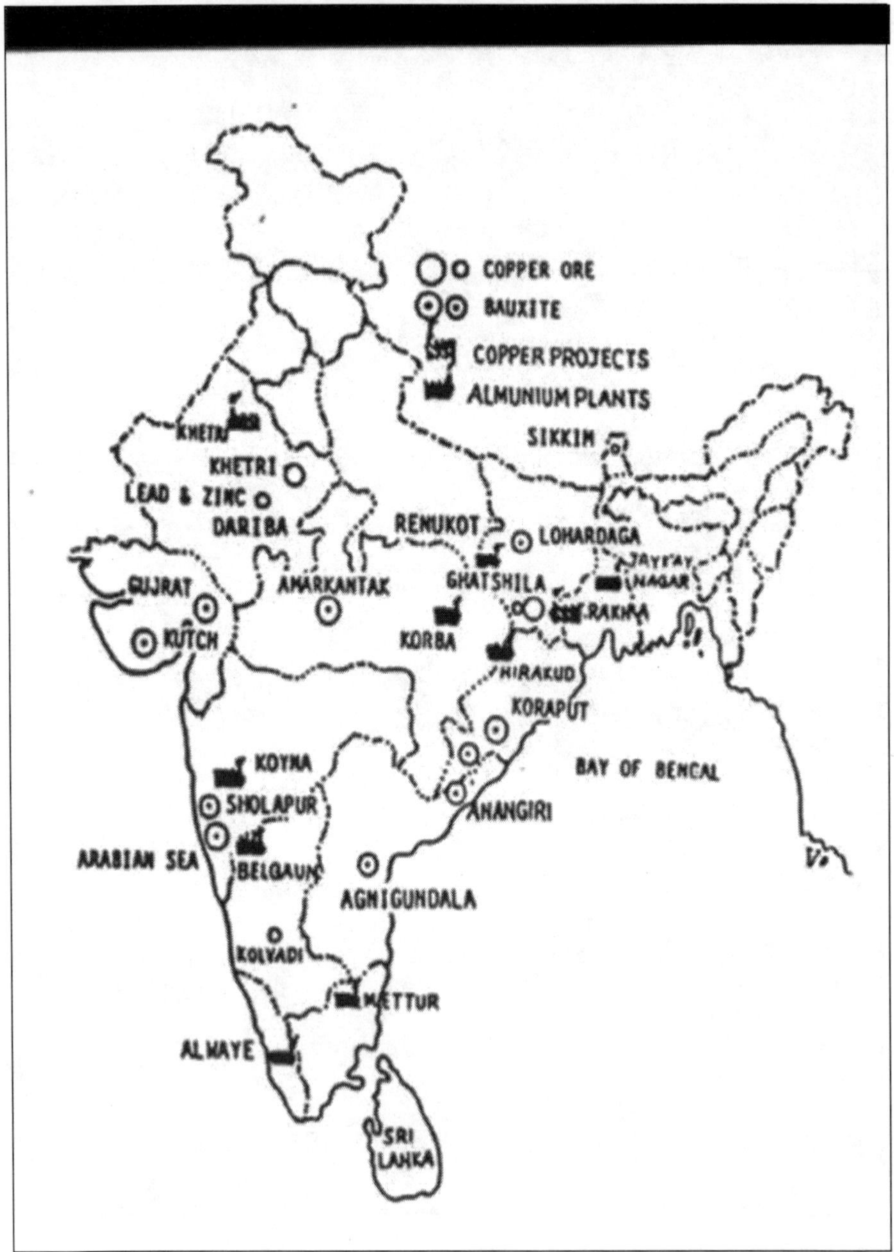

Distribution of Petroleum in the map of India.

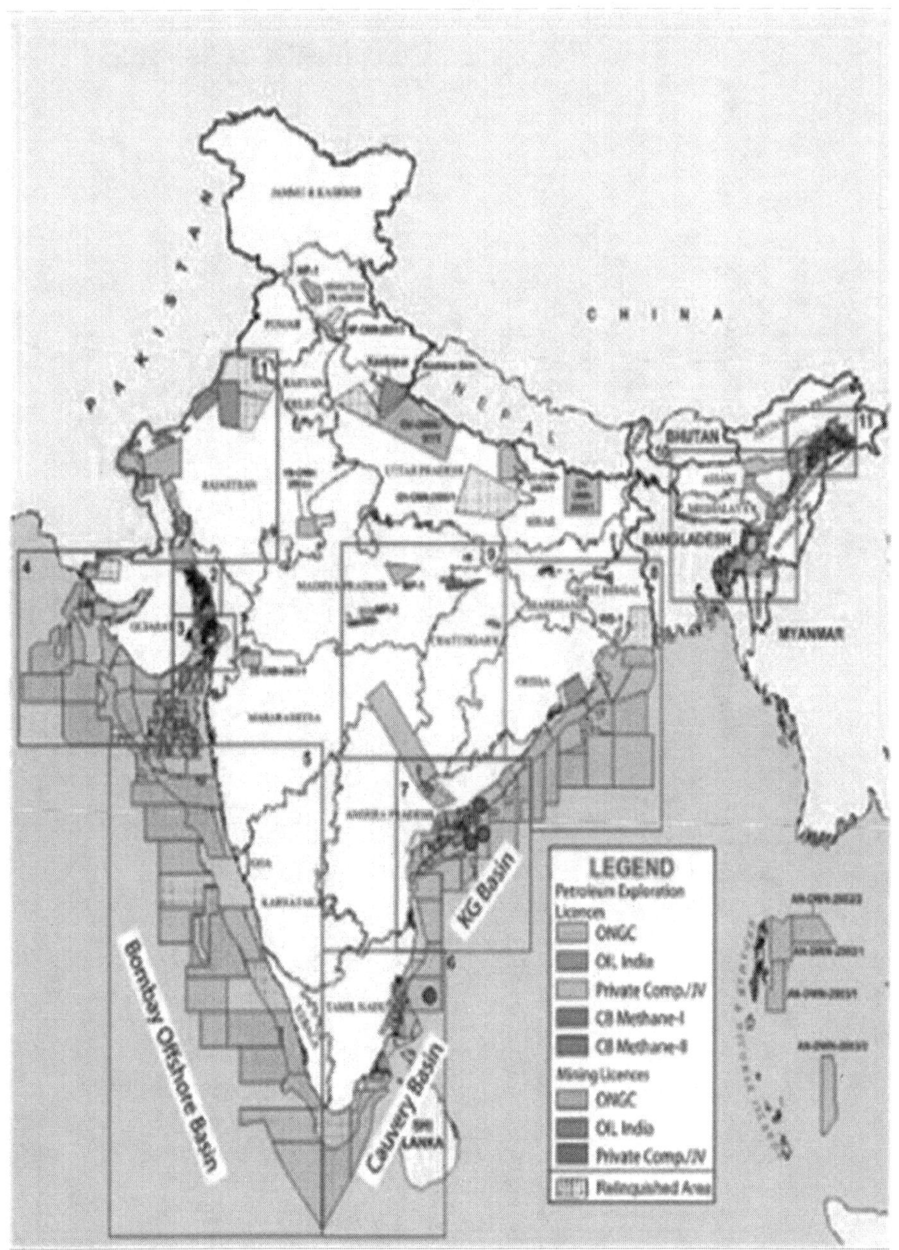

URANIUM OCCURRENCES IN INDIA
New Delhi
Gwalior
Aravallis
Shillong Basin
Son Valley
Singhbhum
Kolkata
Chattisgarh Basin
Mumbai
Bhima Basin
Hyderabad
Gogi
Lambapur - Peddagattu
Cuddapah Basin
Chennai
Operating uranium mines
(Jaduguda, Bhatin, Narwapahar, Turamah)
Prospective uranium mines
Potential uranium belts
Other uranium occurrences

8.2 Ore Reserve Estimation

Once a deposit has been proved by exploration, it becomes more or less a commercial proposition. The more precise the exploration, the more accurate will be reserve estimate. The reserve can be estimated by computing the volume of the ore and converting that volume into tonnage

Tonnage = Volume X Specific Gravity (Bulk Density)

Q = V.D where Q = Tonnage (reserve in tonnes)

V = Volume of the ore (Area X Thickness)

D = Specific gravity or density of raw mineral (bulk density)

8.2.1 Steps of ore reserve calculation

- First the area of influence is calculated around borehole.
- With the help of planimeter, the area of polygon is calculated.
- Thickness and volume are tonnes then determined.
- Now the average thickness and average grade is determined based on following formula:

$$\text{Average Thickness} = \frac{\text{Volume of Material}}{\text{Area of Influence}}$$

$$\text{Average Grade} = \frac{\text{Volume x Assay Product}}{\text{Volume of material}}$$

Exercise 1. Calculate the average grade of data (Table below) supplied by a mine. The cut off grade is 2% Zn. Observation Table:

Sample No.	Length	Assay	Length *Assay	Ore type
Q-1	30	1.6%	-	Zn
Q-2	60	3.2%	192	Zn
Q-3	60	5.1%	306	Zn
	Σ150		Σ 498	

Sample No. Q-1 is rejected because the assay value is less than 2% (below cut off grade).

Solution: Average Grade = Σ (Length *Assay) / Σ Length

As given in the above table

Σ (Length *Assay) = 498 and

Σ Length = 150

Therefore,

Average Grade = 498 /150 = 3.32%

Now to verify the average grade, arithmetic grade can be calculated

Arithmetic Grade = 2.81 + 3.82 / 2 = 3.32%

So the result = Average grade of Zn is 3.32%

Exercise 2. Based upon the data supplied in the table below by mine, calculate the average grade of ore. Cut off grade = 2% Zn,

Observation Table

Sample No.	Length	Assay	Length *Assay	Ore type
Q-1	30	04.0%	120	Zn
Q-2	80	0.2%	16	Zn
Q-3	10	02.8%	28	Zn
Q-4	20	03.0%	60	Zn
	Σ140		Σ 224	

Solution:

Average Assay Value at = Σ (Assay* Width) / MSW

Minimum Stopping Width (MSW)

So Average Assay value = 224 / 140 = 1.6%

Ans. Average Grade of Zn = 1.6%.

Exercise 3. Calculation of average grade by area of influence method

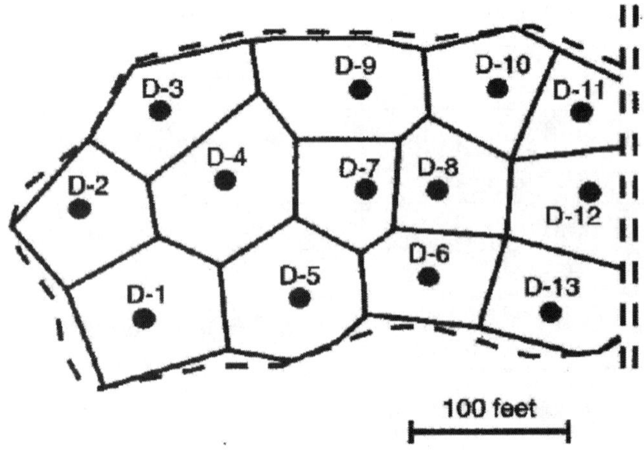

Fig. 8.1 Diamond drill hole plan and polygons for a copper deposit.

Assay Data for Drill Hole D-1. Base on these data determine the average grade of copper ore.

Interval (ft)	Thickness (ft)	Grade % Cu	Grade * Thickness	
0-100	100	0.31		Below Cutoff
100-110	10	0.47	4.70	
110-122	12	0.72	8.75	
122-130	8	0.96	7.68	
130-150	20	1.04	20.80	
150-200	50	0.82	41.00	
200-220	20	0.54	10.80	
220-250	30	0.42	12.60	
250-270	50	0.35		Below Cutoff
	150		106.43	Thickness and Grade* Thickness above cutoff

$$\text{Average Grade Drill Hole D-1} = \frac{\text{Thickness1} * \text{Grade1}}{\text{Thickness1}} = \frac{106.43}{150} = 0.71\%\text{Cu}$$

Exercise 4. Based on the table, given determine the average grade of copper deposit.

Polygon	Area (ft^2)	Thickness (ft)	Volume (ft^3)	Tonnage Factor	Tons	Grade % Cu	Tons * Grade
D-1	5320	150	798,000	12.5	63,640	0.71	45,326
D-2	5300	135	715,500	12.5	57.240	0.66	37,778
D-3	4400	180	792,000	12.5	63.360	0.82	51,955
D-4	5520	175	966,000	12.5	77,280	0.75	57,960
D-5	6800	155	105,400	12.5	84,320	1.00	84,320
D-6	4960	180	892,800	12.5	71,424	0.97	69,281
D-7	4520	250	1,130,000	12.5	90,400	1.21	109,384
D-8	4640	240	1,113,600	12.5	89,088	1.36	121,159
D-9	5840	150	876,000	12.5	70,080	0.93	65,174
D-10	4840	135	653,400	12.5	52,272	0.87	45,476
D-11	3760	120	451,200	12.5	36,096	0.81	29,237
D-12	4270	165	637.200	12.5	50,976	0.75	38,232
D-13	4800	135	648,800	12.5	51,840	0.68	35,251
Total					858;216		790;533

Table 7. Ore Reserves for Copper Deposit

$$\text{Average Grade Deposit} = \frac{\text{Tons} \; * \text{Grade}}{\text{Tons}} = \frac{790.533}{858.216} = 0.92\%\text{Cu} \quad \text{Ans.}$$

8.3 Photogeology and Satellite Remote Sensing

Photogeology: Photogeology is the branch of geology which deals with the use of aerial photographs in geological sciences.

8.3.1 Aerial photography

Aerial photography is the taking of photographs of the ground from an elevated position. Platforms for aerial photography include aircraft, balloons etc. Camera or sensor is triggered remotely with a fixed interval to get photographs of a ground from two different angles. Then these two photographs are studied by stereoscope that produces three dimension effects to this photograph.

8.3.2 Types of Aerial photographs

Aerial photographs are of two types

Vertical aerial photographs: Photographs taken by a camera which is pointing vertically downward is called vertical aerial photographs.

- **Oblique aerial photographs**: Photographs taken with inclined optic axis of camera from the vertical are known as oblique aerial photographs. If the inclination of optic axis of camera is more, it is called high oblique and when less, then low oblique aerial photographs.

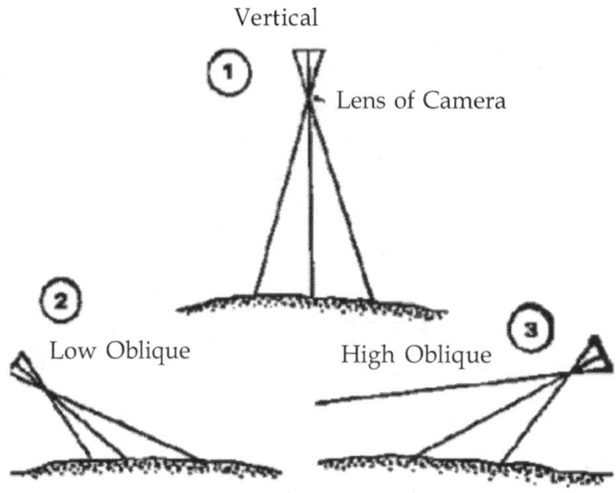

Fig. 8.2 Aerial photograh

8.3.3 Photogrammetry

Photogrammetry can be defined as the art, science, and technology of obtaining reliable information about physical objects and the environment through processes of recording, measuring and interpreting photographic images and patterns of recorded radiant electromagnetic energy and other phenomena.

8.3.4 Photo scale

In map scale, the ration of map distance to ground distance is known as map scale. Photoscale is not uniform nor a regularly changing scale.

The formula of photo scale can be understood by the following geometric calculation as seen in figure.

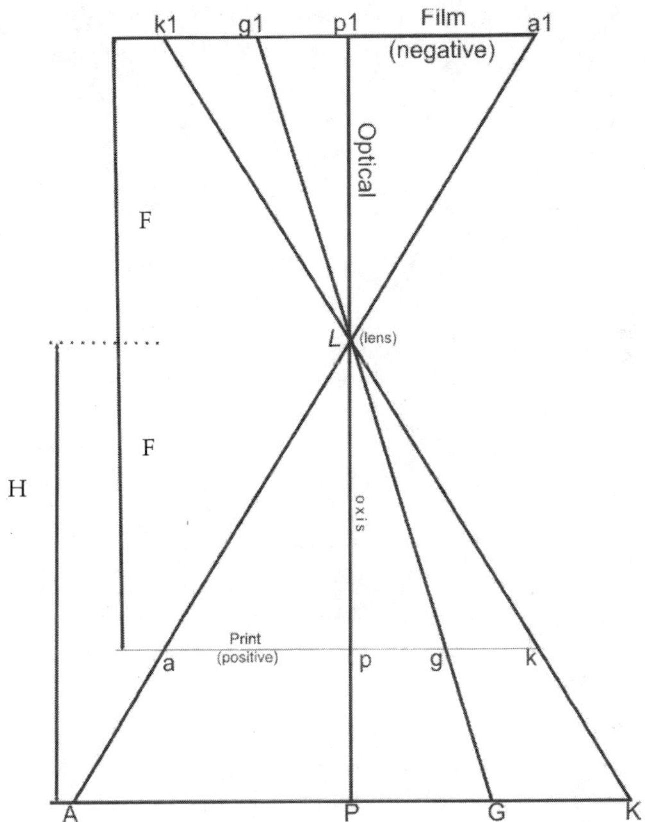

Fig. 8.3 Determination of Photoscale from Vertical Aerial Photograph

Photoscale = Image Distance/Ground Distance

a1.p1/A.P = p1l/PL

= F/H= Focal length/Camera height

So the Photo-scale is the ratio of focal length and camera height.

8.3.5 Stereoscopic vision

Stereoscopy, sometimes called stereoscopic imaging, is a technique used to enable a three dimensional effect, adding an illusion of depth to a flat image. In aerial photography, when two photographs overlap or the same ground area is photographed from two separate position forms a stereo-pair, used for three dimension viewing. Thus obtained a pair of stereoscopic photographs or images can be viewed stereoscopically.

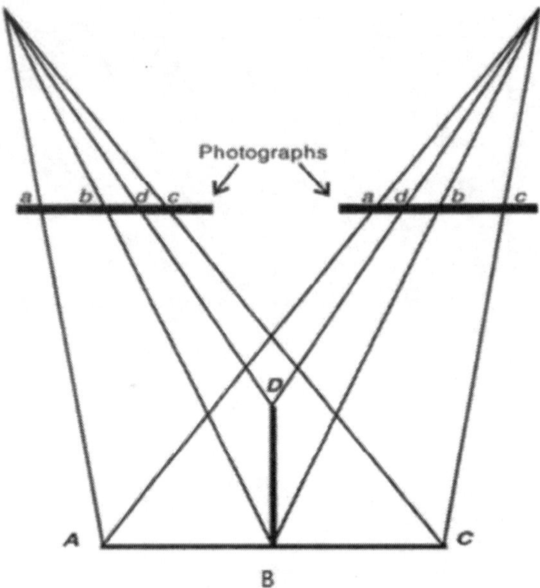

Fig. 8.4a Perception of relief from two aerial photographs.

8.3.6 Stereoscope

A stereoscope is used in conjunction with two aerial photographs taken from two different positions of the same area, (known as a stereo-pair) to produce a 3-D image. There are two types of stereoscopes: lens (or pocket) stereoscope and mirror stereoscope. Lens (or pocket) stereoscope has a limited view and therefore restricts the area that can be inspected where as in mirror stereoscope has wide view and enables a much larger area to be viewed on the stereo-pair.

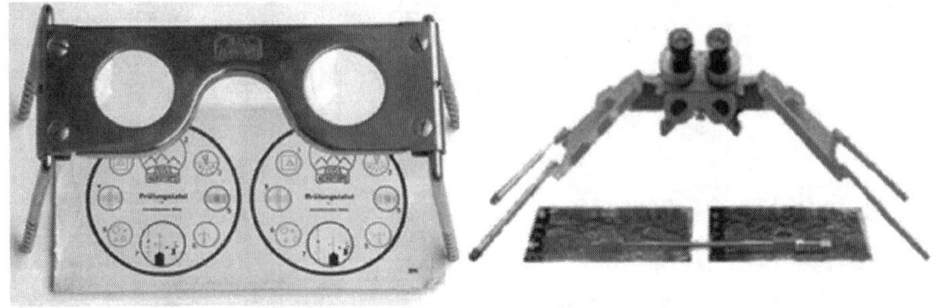

A. Pocket B. Mirror stereoscope

Fig. 8.4b Types of Stereoscope

8.3.7 Identification and interpretation of aerial photographs

For the identification and interpretation of aerial photographs, the following basic points must be considered:

- Identification of representative structure present in nature can be used as a key to recognize geological features in aerial stereo pair.
- For the identification of features, student must require patience, judgement and ability to evaluate the significance of different types of information.
- Many geological features can be identified by the Shape and Form like Dome, Cinder cone, Sand dunes, Glacial deposits, Karst topography, Alluvial fan, Meandering of river, Ox bow lake, Caldera lake. Plunging fold etc.
- Aerial photographs are used in photogeological studies. All colors are reduced to various grades of gray tones. Different gray tones are termed as tones, light or dark. Texture of an aerial photograph is the quality by which we can identify it with other feature having same gray tone, Different rocks weather to different colour than fresh exposed outcrop. Photographic tone and texture are very important to identify different rock types.
- Drainage pattern exhibit different nature in different rock types or structures, and is very important to identify Igneous sedimentary and metamorphic rocks. Ex. Dendritic pattern of drainage is indicative of igneous rock particularly granite rocks. Drainage pattern can also be helpful in interpreting structures like dome, basin, joints, folds, plunging folds.

8.4 Practical Exercises on Identification and Interpretation of Aerial Photographs by Stereoscope

A. Action of River: Meandering (Fig. 8.5)

i. The given stereo-pair shows action of River. A broad flood plain has created by a stream.

ii. In reduced velocity, the river moves in Zigzag manner. This is called meandering of river.

iii. In this stage, complex branching system of river is developed. The river flows in an almost uniform gradient. The drop in gradient reduces the velocity which in turn decreases its erosive power.

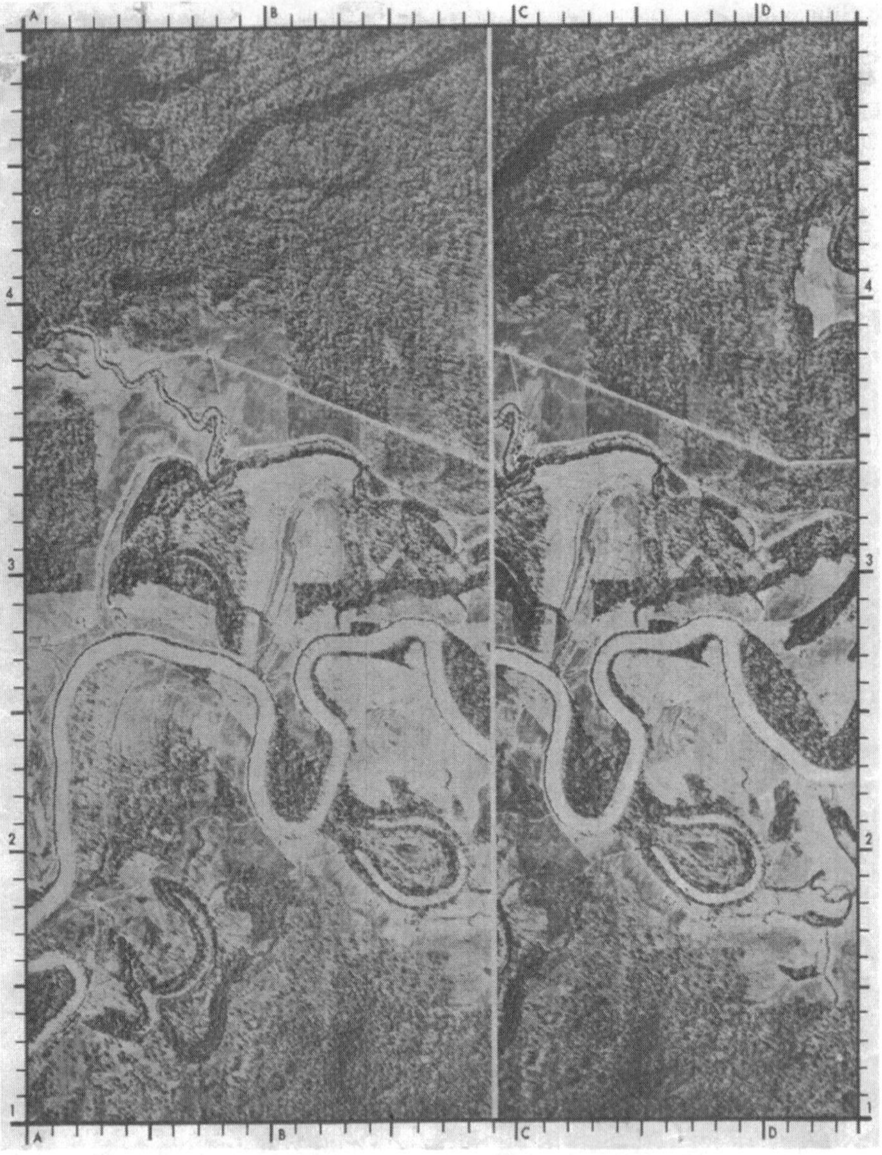

Fig. 8.5 Stereopair showing Meandering of River and formation of Ox-Bow lake

iv. At places meanders leaving the older channel as horse shoe shaped structure called Ox-Bow lake.

v. At the top left of stereopair, the dark tone denotes the forested area and a dark band represents abandoned channel.

B. Karst Topography: (Fig. 8.6)

i. The given stereopair shows Karst topography.

ii. In limestone country, warm water of tropical regions dissolves limestone much rapidly than cool water of a temperate region.

iii. Due to dissolution in limestone, small and deep holes are formed and are called Sink holes.

iv. So in limestone area, the rainwater does not form any drain but sink through these holes and becomes a part of groundwater.

v. At places sink holes joined by other sink holes and form natural bridges. The total topography in limestone region is called Karst topography (mottled appearance at the top).

C. Valley Glacier: (Fig. 8.7)

i. It is clear from the stereo pair that the steep sided U shaped valley is formed at the centre and is flowing towards the west side.

ii. Dark band at the border are lateral moraines. At the middle it is called medial morain.

iii. The tip of Glacier contains transported debris, that is known as terminal morain.

iv. The U shaped valley is known as Hanging valley.

v. Based on characters described above, the stereo pair represent a Valley Glacier.

D. Cinder Cone: (Fig. 8.8)

i. The given stereo pair clearly shows an active volcanic cone.

ii. The black cinder is basaltic in composition and it is flowing in the natural slope direction (west).

iii. There is another weathered cone in the north of this symmetrical cone.

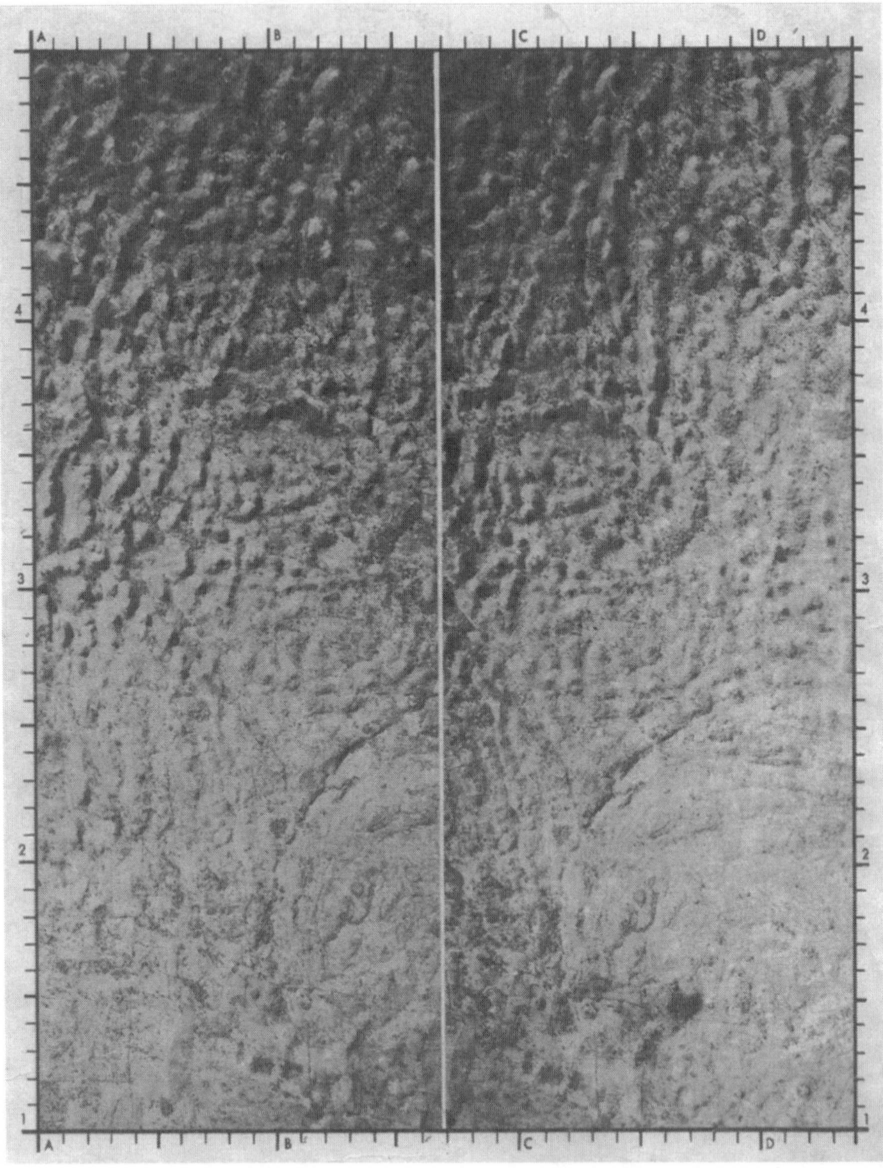

Fig. 8.6 Stereopair showing Karst topography in limestone country

Fig. 8.7 Stereopair showing Valley Glacier.

Fig. 8.8 Stereopair showing Cinder Cone Volcanic activity

E. Caldera lake (Fig. 8.9)

i. The given stereo pair is a big Caldera lake (dark colored) in the west side of the photograph.
ii. Earlier, the area was a part of active volcano and big cinder cone was developed around it.
iii. The remnants of volcanic plug are seen at the lower part of the Caldera.
iv. The Caldera was created by the collapse of Cinder cone, its border was intact and the rain water collected at the centre and formed a Caldera lake.

F. Action of wind, Formation of Dunes: (Fig 8.10)

i. The given stereo pair represents action of wind in a desert area.
ii. The desert floor shows a fine display of sand dunes.
iii. Here the crescent shaped dunes are called Barchan dunes.
iv. The dune looks like English letter C and the concave direction of it shows the direction of wind.

8.5 Satellite Remote Sensing

Remote sensing is broadly defined as the collecting and interpreting about an object without being in contact with the object. Aircrafts and satellites are the common plateform for Remote Sensing data collection. With the help of electromagnetic radiation, information is taken about objects on the Earth's land surface, oceans, and atmosphere through the sensors attached to the satellites. As mentioned earlier, the aerial photographs taken from aircrafts initially proved useful in geological mapping. Later images produced by satellite imaging system from multispectral sensors showed tremendous potential as a important source of application in various branches of geology like geomorphology, mineral exploration, structural and lithological mapping projects. There is a limitation in map prepared by remote sensing, where the area is forest covered, a field validation is compulsory to confirm the structure (Fig. 8.11).

1. Energy Source or Illumination Sun (A),
2. Radiation and the Atmosphere (B),
3. Interaction with the Object (C),
4. Recording of Energy by the Sensor (D),
5. Transmission, Reception and Processing (E),
6. Interpretation and Analysis (F),
7. Application (G)

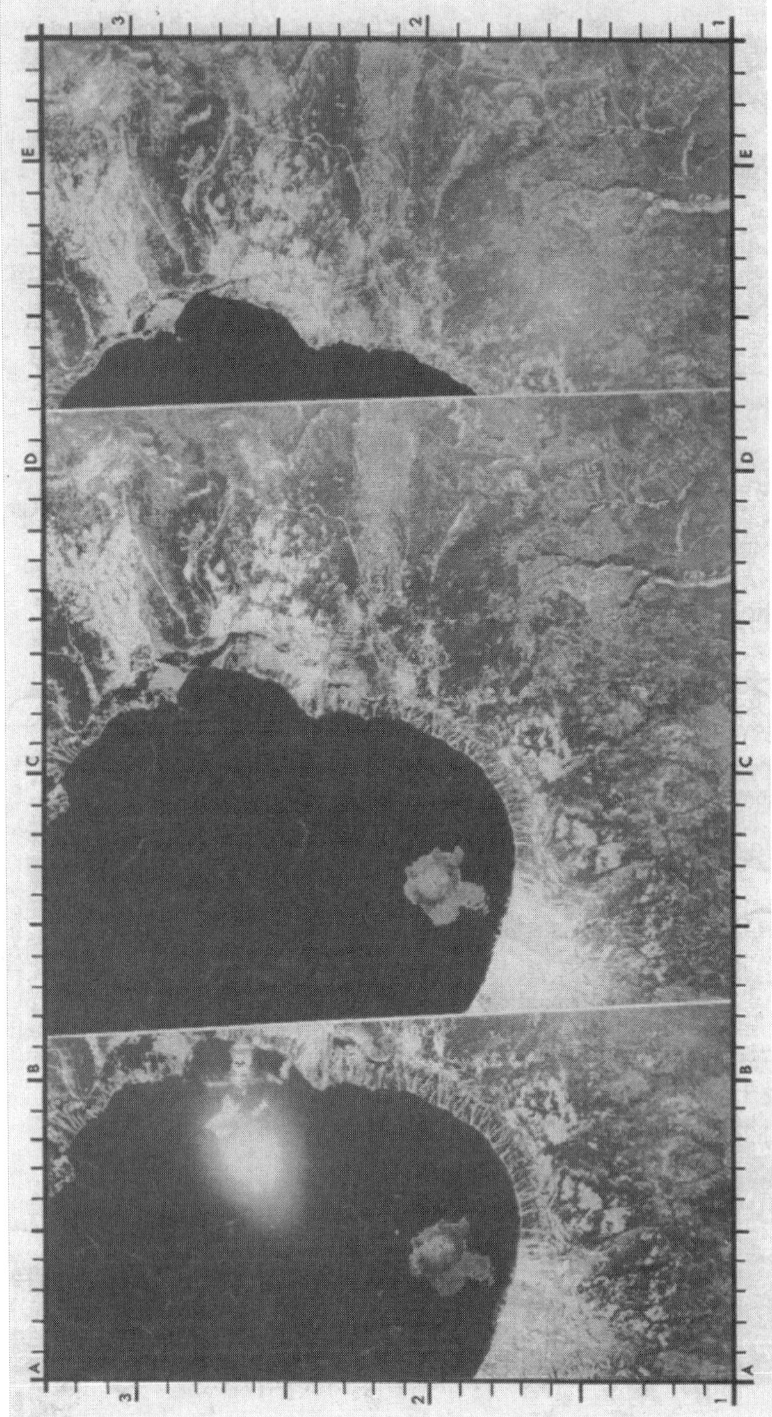

Fig. 8.9 Stereopair showing Caldera lake

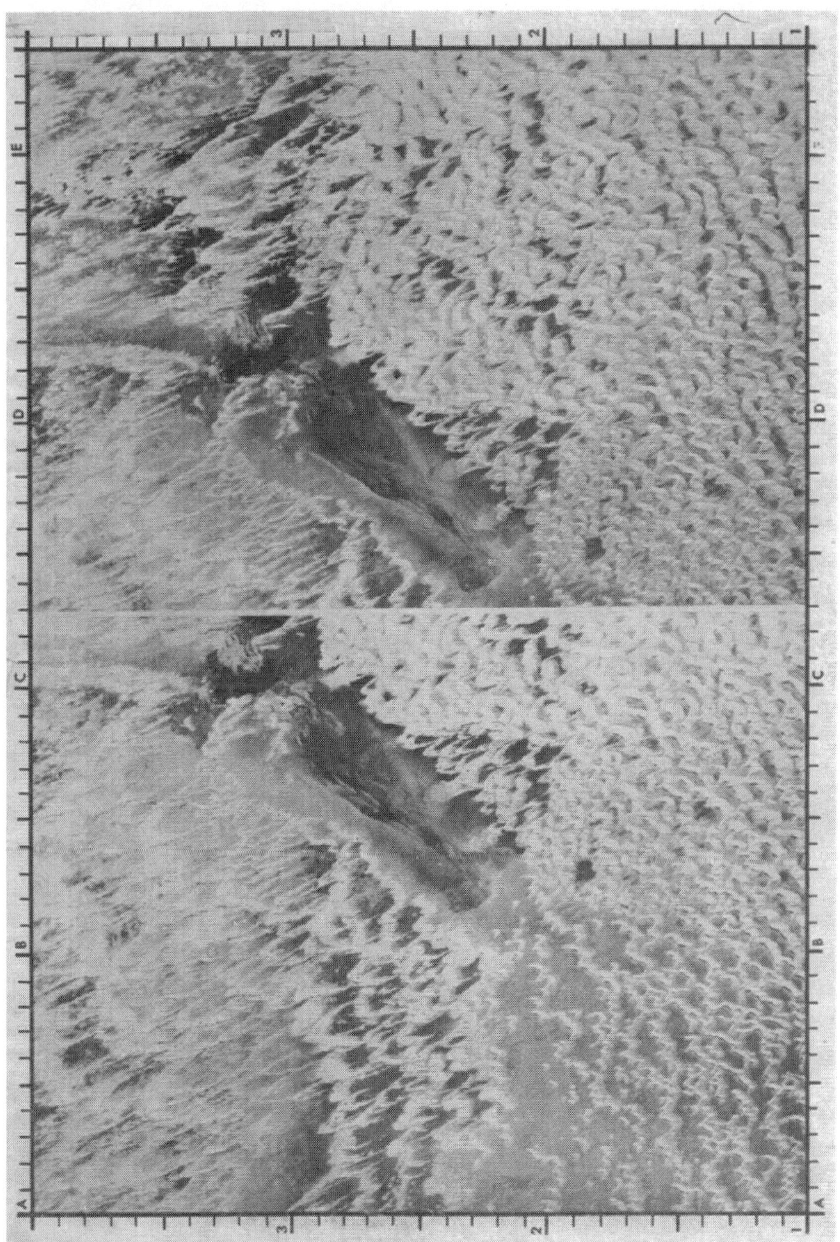

Fig. 8.10 Stereopair showing formation of Dunes in a Desert area.

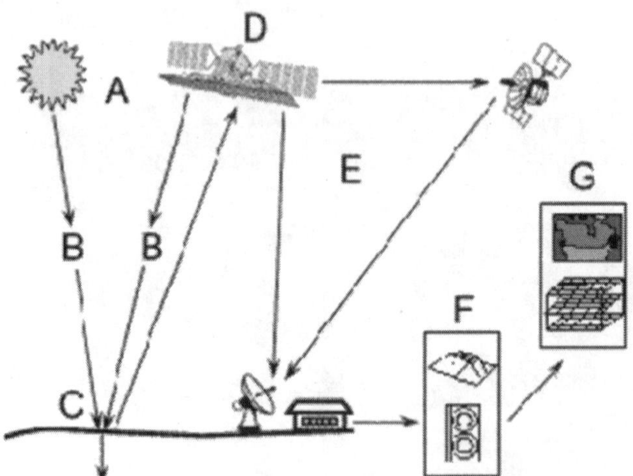

1. Energy Source or Illumination Sun (A),
2. Radiation and the Atmosphere(B),
3. Interaction with the Object (C),
4. Recording of Energy by the Sensor (D),
5. Transmission, Reception and Processing (E),
6. Interpretation and Analysis (F),
7. Application (G)

Fig. 8.11 Elements of remote Sensing

8.5.1 Electromagnetic radiation

Electromagnetic Radiation (EMR) is the radiant energy released by certain electromagnetic processes. Visible light is one type of electromagnetic radiation; other familiar forms are invisible to the human eye, such as radio waves, infrared light & X-rays etc. (Fig. 8.12).

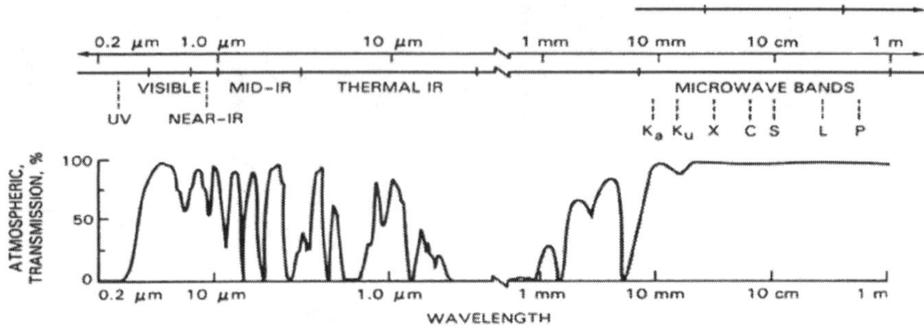

The Electromagnetic (EM) Spectrum

Fig. 8.12 Electromagnetic Radiation (EMR)

8.5.2 Sensors

Passive Sensor: A passive sensor is a microwave instrument designed to receive and to measure natural emissions produced by constituents of the Earth's surface and its atmosphere. The power measured by passive sensors is a function of the surface composition, physical temperature, surface roughness, and other physical characteristics of the Earth (Fig. 8.13 A).

- *Active Sensor:* An active sensor is a radar instrument used for measuring signals transmitted by the sensor that were reflected, refracted or scattered by the Earth's surface or its atmosphere. Space borne active sensors have a variety of applications related to meteorology and observation of the Earth's surface and atmosphere. (Fig. 8.13 B).

Fig. 8.13A Passive Sensor

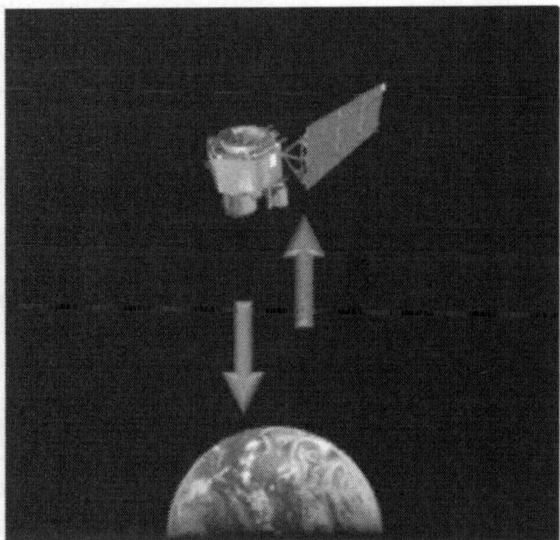

Fig. 8.13B Active Sensor

8.5.3 Types of satellite

Sun-Synchronous orbits

Meteorological satellites are often placed in a sun-synchronous or helio synchronous orbit. These satellites are in polar orbits. The orbits are designed so that the satellite's orientation is fixed relative to the Sun throughout the year, allowing very accurate weather predictions to be made. Most meteorological (Sun-synchronous) satellites orbit the Earth 15 to 16 times per day at the altitude of approximate 600 to 800 km above msl.

Geostationary satellites

The majority of communications satellites are in fact geostationary satellites. Geostationary satellites take 24 hours to complete a rotation. However, geostationary satellites are positioned directly over the equator and their path follows the equatorial plane of the Earth. As a result geostationary satellites don't move North or South during the day and are permanently fixed above one point on the equator of the Earth. Most video or T.V. communications systems use geostationary satellites. Geostationary satellites are typically orbiting at 35,788 km (22,238 miles) above the surface of the planet (42,000 km from its centre).

8.5.4 Image interpretation

Human eyes have very limited capacity to visualize in the field. Remote sensing techniques provide a bird's eye view, and can be helpful in determining and interpreting many geological structures. A synoptic view helps in visualizing terrain as a whole. For image interpretation, a broad geological knowledge about the area is very essential. The geological map is prepared by this method can be used in mineral exploration, engineering geology related projects, environmental geology etc.

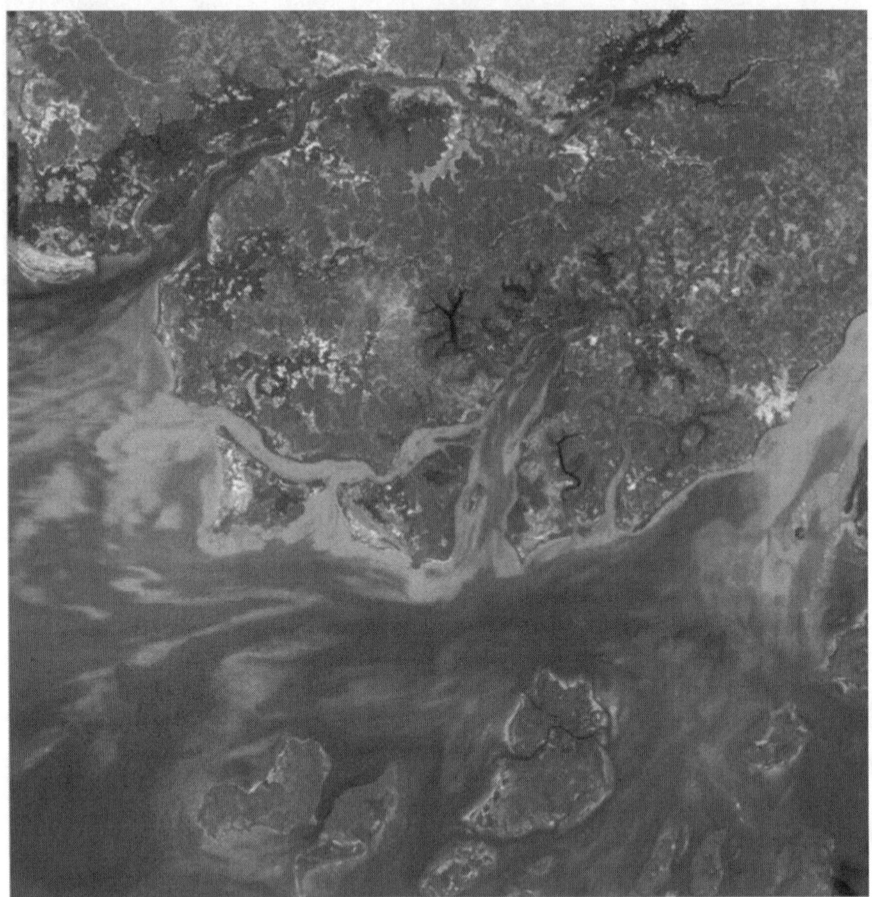

Fig. 8.14 Landsat Satellite Image (*See colour version on page 234*)

8.5.5 Image elements

Following elements are helpful in satellite image identification and interpretation:

- **Tone/colour**: Tone is directly related to reflectance of light from terrain. For example dry sand reflects high % of incident rays and gives bright tone to the image while wet sand gives dark tone. In false colour composite (FCC) images water bodies looks blue while dry sand in white tone.
- **Texture** :Texture is dependent on scale, as the scale of photograph is reduced, the texture of given object becomes progressively finer and eventually disappear.
- **Pattern**: The pattern relates to the spatial arrangement of the object. The repetition of certain general forms, is characteristic of many objects, and gives objects a pattern. Example: Inter-bedded sedimentary rock typically gives an alternating tonal pattern, gives their identity.
- **Shape** : Shape relates to the general form, configuration or outline of an individual object. Shape is one of the most important single factors for recognizing objects from image. Ex. Railway line is usually distinguished from a highway.
- **Size**: The size of an object can be important tool for the identification. If the size is not evaluated properly, object can't be interpreted, but valuable information can be derived from the shadow of the object. But it is better interpreted in aerial photographs.
- **Association**: Association is best for identification of landforms such as terraces, meanders, ox bow lakes, abandoned channels etc.

Based on these image elements, one can interpret satellite images. Unlike aerial photo (stereopairs), satellite images are being studied only with single photo. We can order tailor made images in different scale from National Remote Sensing Agency, Hyderabad.

❑❑❑

Colour Plates

Chapter 4: Minerals and Rocks

Diopside
Asbestos
Actinolite
Hornblende
Biotite
Muscovite
Quartz Crystals
Quartz

Plagioclase Felspar
Zeolite
Orthoclase
Dolomite
Dolimite rhombic crystals
Fluorite cubic crystals (light blue)
Fluorite (Dark crystals)
Pyrolusite

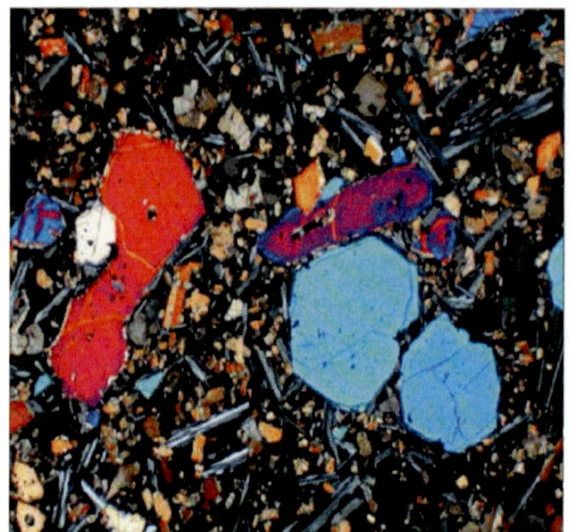

Olivine in Thin section, showing zonal structure crossed polar

Olivine in Gabbro rock crossed polar section

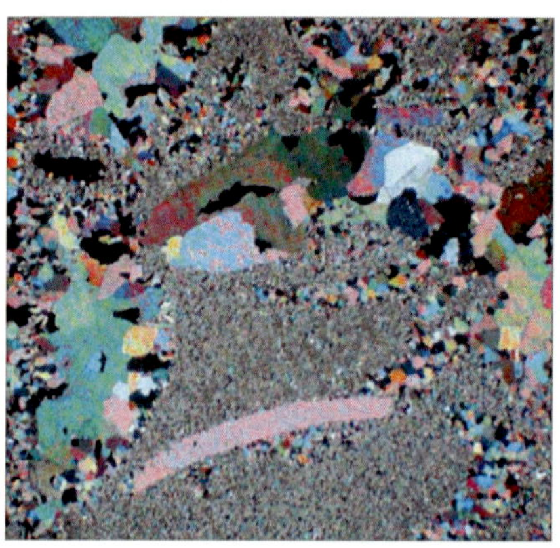

Calcite spar and micrite

Rhombohedral crystals of Dolomite with calcite

Quartz showing undulose extinction (Wavy)

Microcline showing characteristic cross hatching

Plagioclase feldspar in thin section

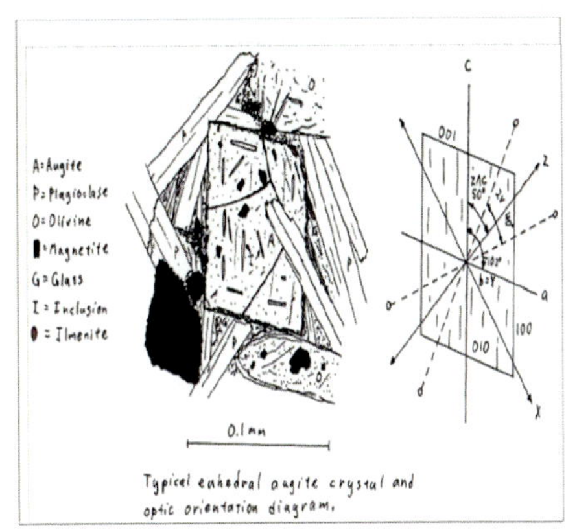

Typical euhedral augite crystal and optic orientation diagram.

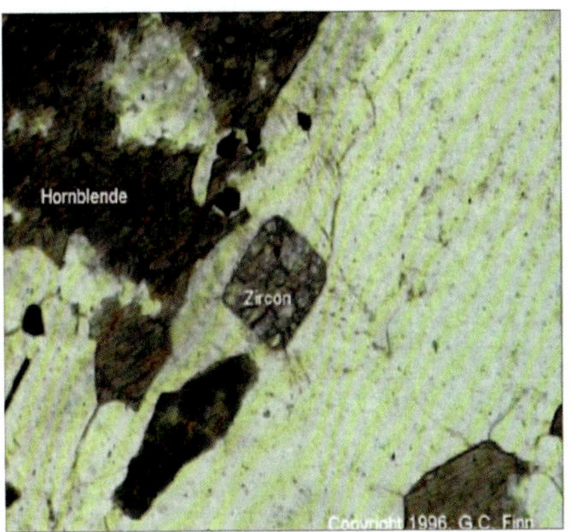

Zircon

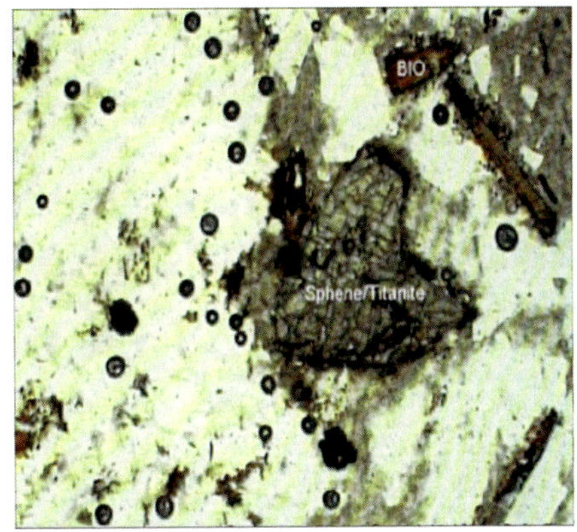

Sphene

Euhedral Garnet in schist

Hornblende
Green hornblende in a diorite. Pleochroism ranges from light yellow-green to bluish-green to brownish-green.

Kyanite in thin section

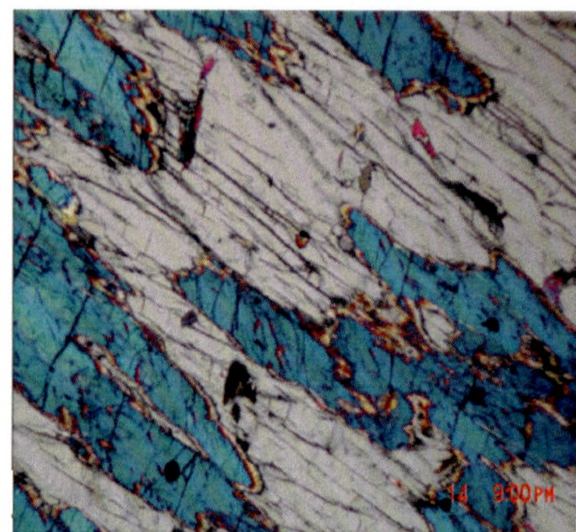

Thin section, crossed polars: Andalusite (gray) with sillimanite (blue)

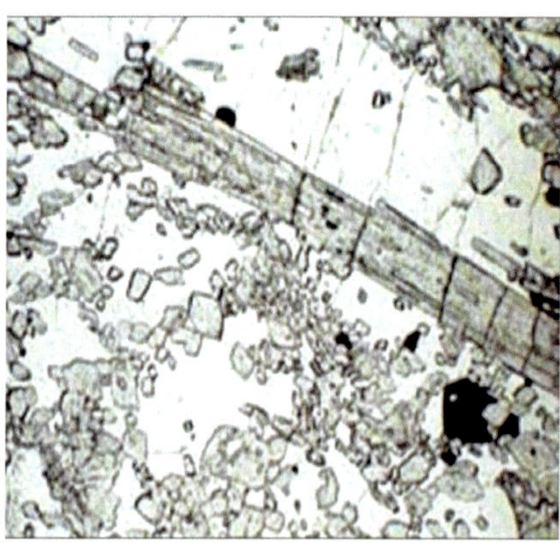

Sillimanite in uncross condition

Sillimanite in crossed Nicol condition

Prismatic crystal of tourmaline in PPL (above) and XPL (Below).

Corundum in thin section

Beryl in plane polarized light

Augite in crossed polar

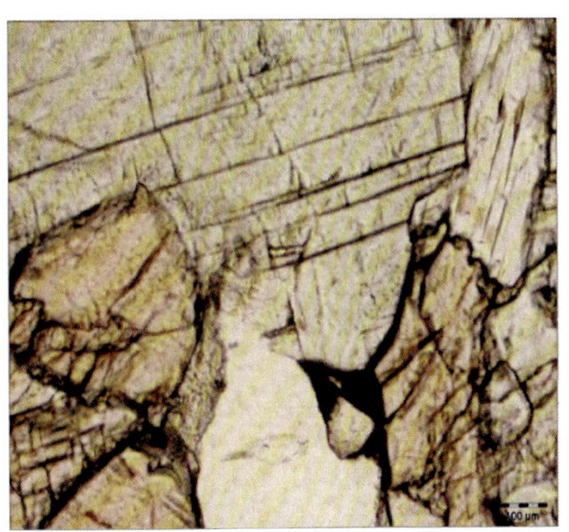

Photomicrograph of diopside in plane polarized light. Diopside is the pale green mineral at the top of the photo; it is surrounded by reddish garnets

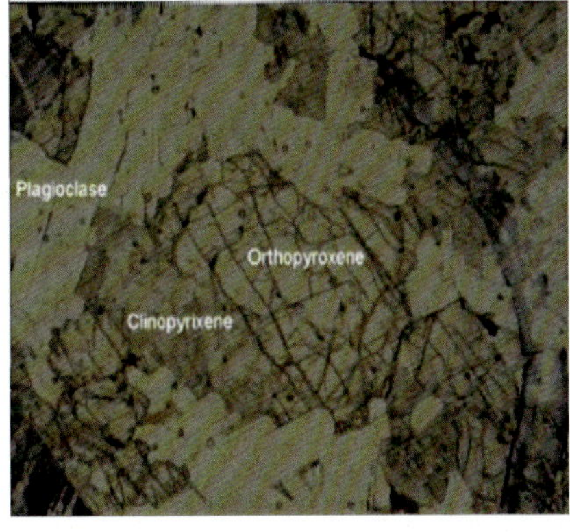

Pyroxene in thin section Ortho and Clino

Biotite
Mineral biotite shows pleochroism in ppl.it changes color from dark brown to black when the thin section is rotated

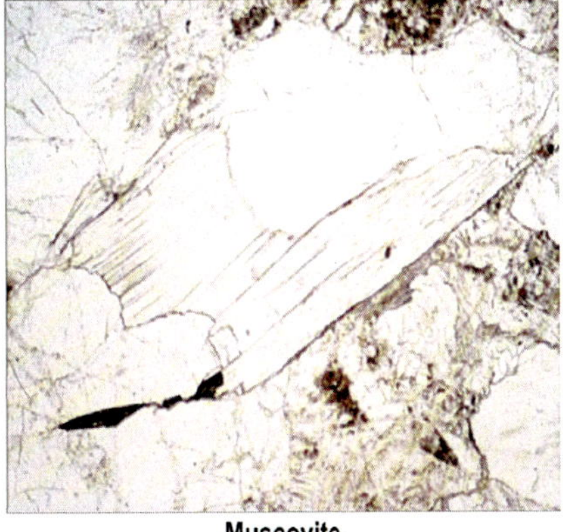

Muscovite,
Muscovite is generally colorless with good cleavage. Pale brown radiation halos can sometimes be visible.

Granite

Rhyolite

Obsidian

Pumice

Scoria

Diorite

Gabbro

Pegmatite

Andesite

Basalt

Sedimentary Rocks

Clastic rocks

Shale

Sandstone

Sandstone

Siltstone

Salt

Dolomite

Coal

Gneiss

Phyllite

Schist

Slate

Marble

Amphibolite

Hornfels

Quartzite

Granite in thin section

Gabbro in thin section

Basalt in thin section

Sandstone in thin section

Dolomitic limestone in thin section

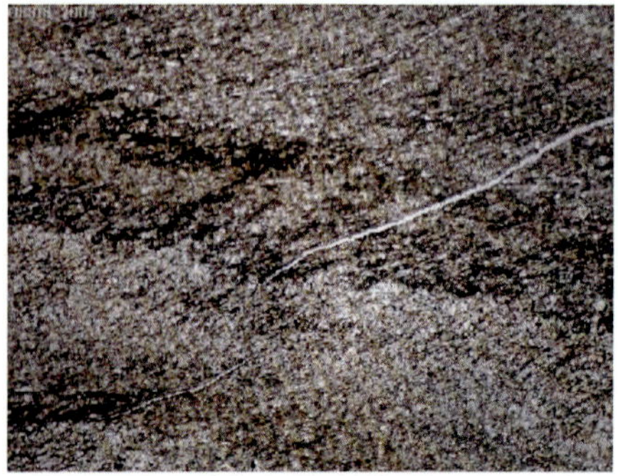

Slate in thin section

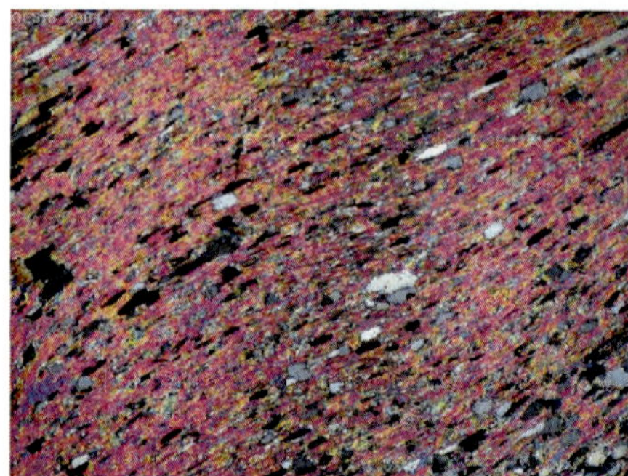

Phyllite in thin section

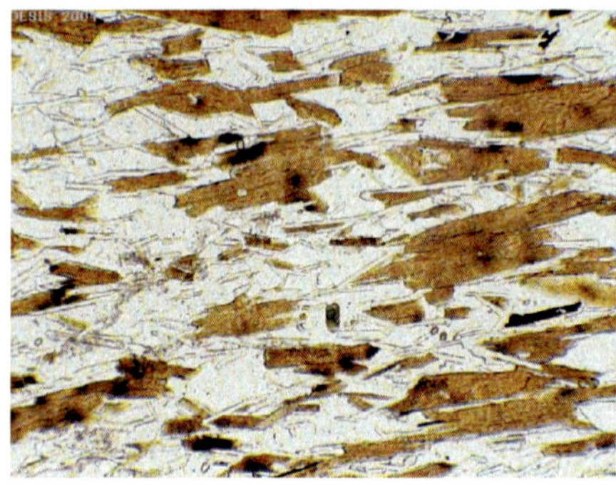

Mica Schist in thin section

Gneiss in thin section

Marble in thin section

Chapter 5: Palaeontology

Phylum Protozoa Nummulites

Phylum Coelenterata Calceola

Phylum Coelenterata Zafrentis

Monograptus (Graptolites)

Diplograptus (Graptolites)

Brachiopod Morphology

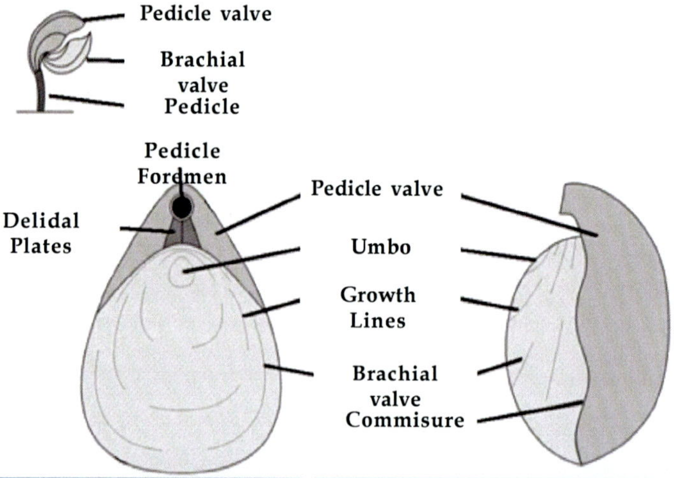

Pedicle valve
Brachial valve
Pedicle

Pedicle Foremen
Delidal Plates
Pedicle valve
Umbo
Growth Lines
Brachial valve
Commisure

www.fossilshells.nl

Productus

Spirifer

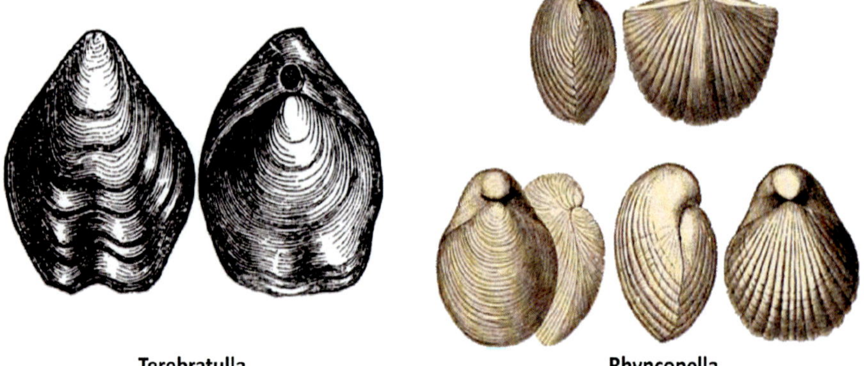

Terebratulla

Rhynconella

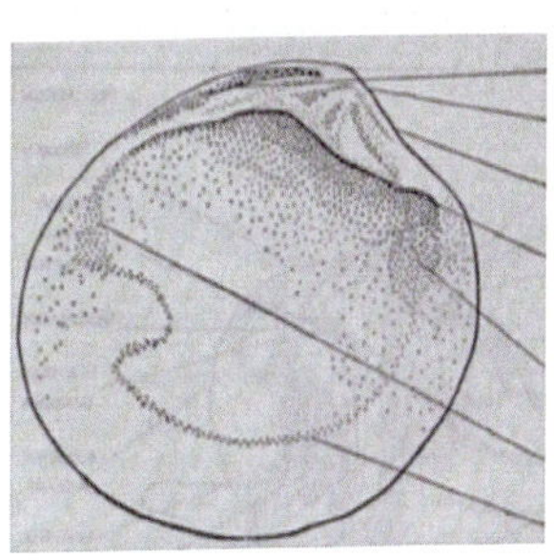

Lamellibranchia Morphology

Venus

Cardita

Exogyra

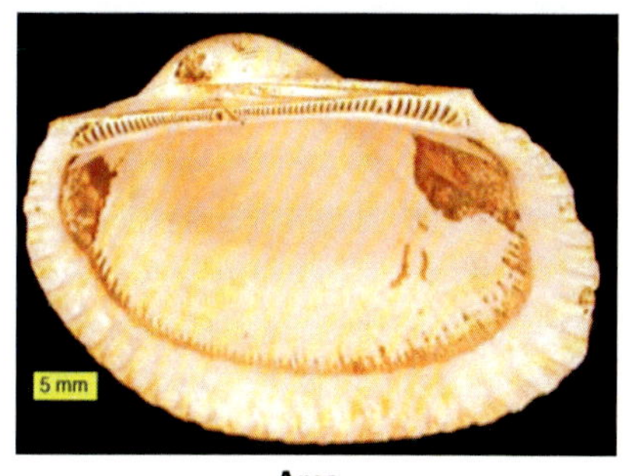

Arca

Ostrea

Gryphaea

Pecten

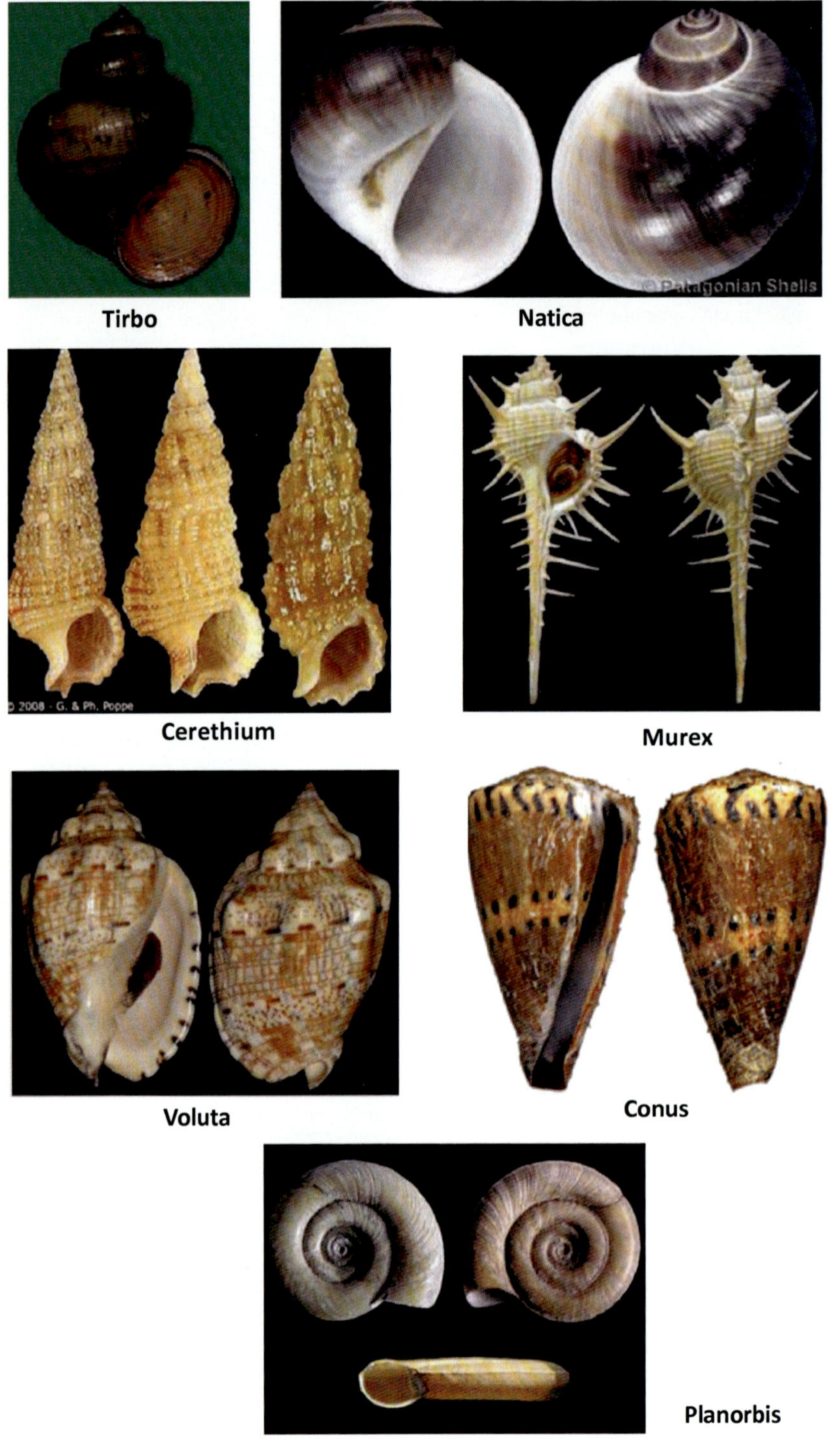

Tirbo **Natica**

Cerethium **Murex**

Voluta **Conus**

Planorbis

Nautilus **Goniatite** **Ceretites**

Scaphites **Acanthoceras**

Perisphinctus

Evolution of Ammonitic suture line with age.

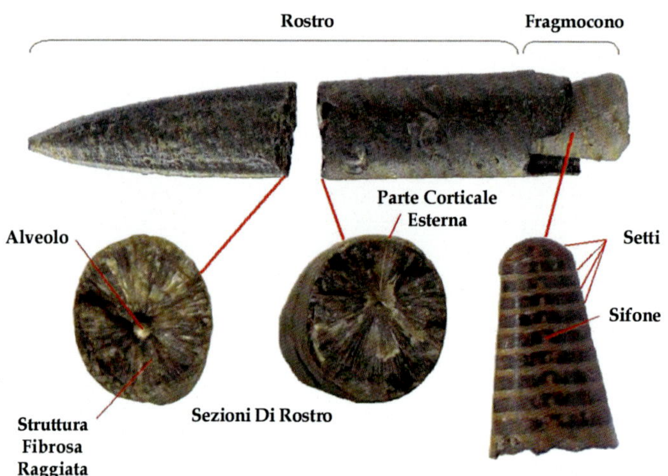

Balemnites

Orthoceras

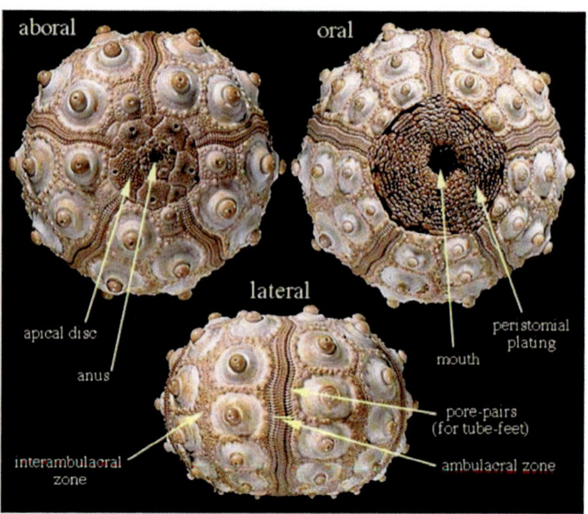

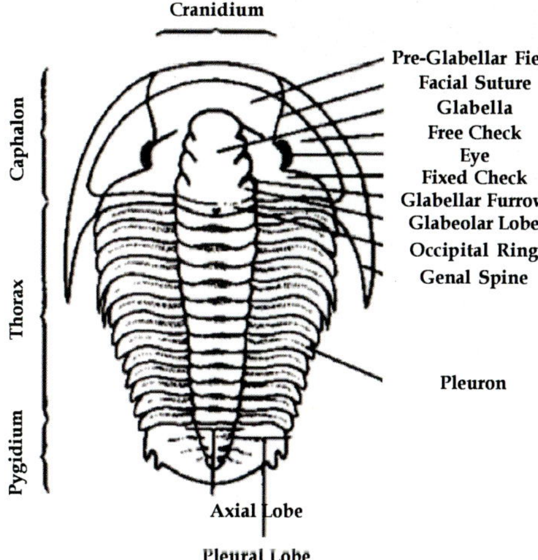

Paradoxide

Phacops

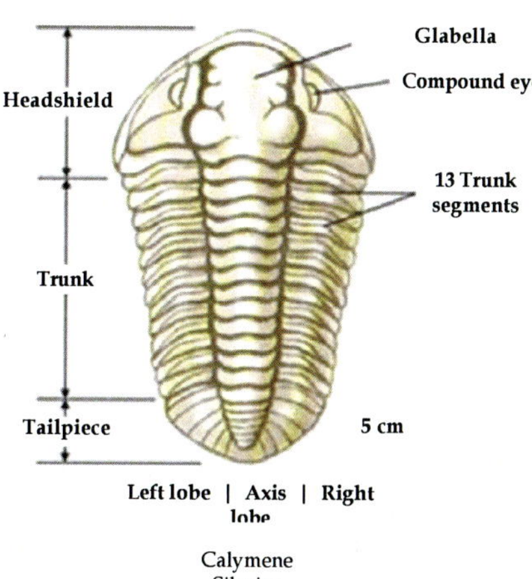

Calymene
Silurian

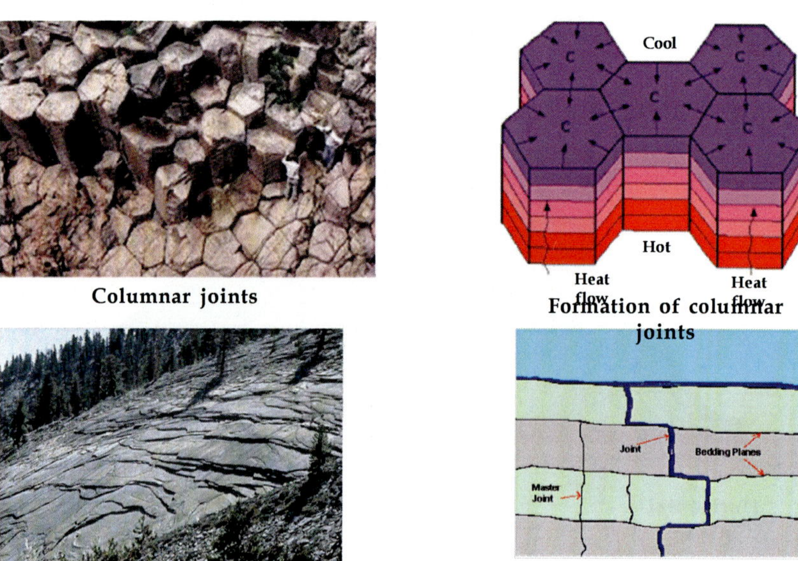

Columnar joints

Formation of columnar joints

Sheet joints

Bedding plane joints

Fig. 7.6 Types of Joints

Chapter 8: Economic Geology, Ore Mineralogy, Remote Sensing and Photogeology

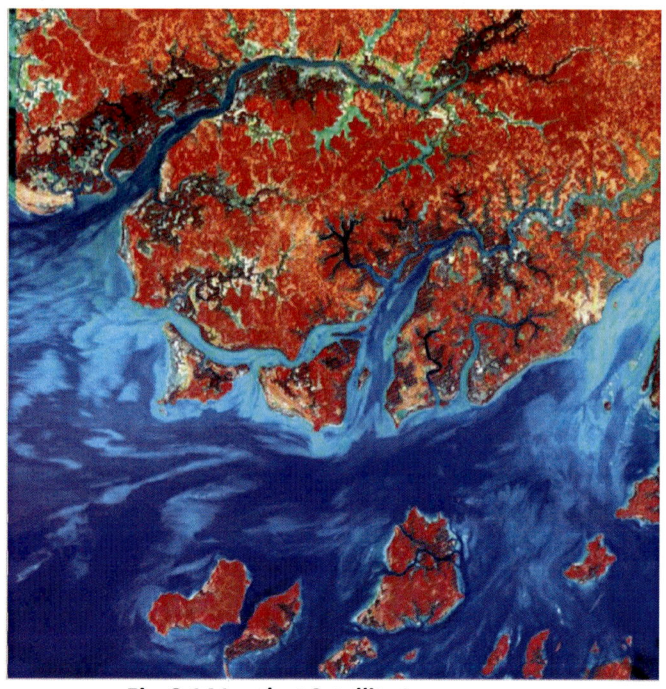

Fig. 8.14 Landsat Satellite Image

Acknowledgement

Getting this book into your hands was a joint effort. I acknowledge the contributions of my colleagues, students and reviewers who helped in deepening my learning for the book **"Geology: Principles and Practical Manual"**. Sincere thanks to all the reviewers who helped me to improve the book with their incisive comments. And finally, a very special thanks to my wife Dr Arti and lovely children Abhishek and Anilabh, for being a constant source of motivation and encouragement. I also acknowledge the following books and internet links that helped me in completing this huge task:

- The Earths Dynamic Surface by K Siddharth
- Physical Geology by A Holms
- A Text Book of Geology by P.K. Mukherjee
- Rutley's Elements of Mineralogy by C.D. Gribble
- An Introduction to Geological and Structural Maps by G.M. Bennison and K.A. Moseley
- Simple Geological Structures by J.I. Platt and J Challinor
- Manual of Geological Maps by N.W. Gokhale
- Introduction to the Study of Geological Maps by A.K. Roy
- A Series of Elementary Exercises upon Geologic Maps by J.I. Platt.
- Principles of Petrology by G.W. Tyrrell
- Mineralogy by Berry, Mason, Dietrich.
- Invertibrate Palaeontology by Henry Wood
- Palaeontology by Anantaraman S.

❒ Fundamentals of Historical Geology by	Ravindra Kumar
❒ Economic Mineral Deposits by	Bateman A. M. & M.L. Jensen
❒ Dana's Text Book of Mineralogy by	William E. Ford
❒ Ore Deposits of India by	KVGK Gokhale and Rao
❒ Ore Microscopy and Ore Petrology by	Craig J.R. and Voughan D.J.
❒ Photogeology by	Miller V.C. and Miller C.F.
❒ Aerial Stereo Photographs by	Waniess H.R.
❒ Image Interpretation in Geology by	Drury S.A,
❒ Remote Sensing: Principles and Interpretation by	Sabins F.F.
❒ Principles and Applications of Photogeology by	Pandey S.N.
❒ Google Internet Links related to Geology	Geology.com and like many more.
	Geoman's home page
❒ Identification Tables for Minerals in Thin Section	Saggerson E.P. Longman, London and NewYork

National Institute of Technology
MINING ENGINEERING SYLLABUS
Name of Subject: Geology-I | Semester: B. Tech. III Sem.
Practical Exercises/ Experiments

List of Experiments

1. Megascopic Description of Rock Forming Minerals.
2. Megascopic Description of important Igneous, Sedimentary, Metamorphic Rocks.
3. Basic Concept of Contours, Attitude of Beds, Width of Outcrop, True and Apparent Dips, Rules of V`s.
4. Study of Geological Maps and Preparation of Cross Sections.

National Institute of Technology
MINING ENGINEERING SYLLABUS
Name of Subject: Geology | Semester: B. Tech. IV Sem.
Practical Exercises/ Experiments

List of Experiments

1. Megascopic description and distribution of Ore Forming Minerals and industrial Minerals. Experiment
2. Study of plant Fossils Experiment
3. Study of Advance Geological Maps and preparation of Cross Sections. Experiment
4. Plotting of important mine-locations on map of India.

METALLURGY ENGINEERING SYLLABUS
Name of Subject: Minerals and Ores Beneficiation
B.Tech. III Sem. | Practical Exercises/ Experiments

1. Mineralogy: Physical Properties of Minerals; Classification of various Rock Forming Minerals, Introduction and Preliminary study of Principle Rock Forming Mineral Groups; Quartz, Feldspar, Mica, Pyroxenes, Amphibole, Garnet, Feldspathoids.
2. Mode of Occurrence, Origin, Distribution, Association and Industrial uses of important Metallic (Al, Au, Cu, Cr, Fe, Mn, Sn, Pb and Zn).

Pt. Ravishankar Shukla University, Raipur
Practical Syllabus | B. Sc. Ist year

Laboratory Work

1. Study & drawing of block diagrams of important geomorphological models. Reading topographical Maps, Interpretation of landform and drainage pattern from topographic maps. -5 Mks
2. Exercises on structural geology problems: completion of outcrops, Drawing and interpretation of cross-sections through elementary representative geological structures. - 6 Mks
3. Study of elements of symmetry of at least one representative crystal of normal classes of each crystal system. Study of physical properties of important minerals in hand specimens. -7 Mks
4. Study of optical characters of important rock forming minerals using polarizing microscope. -4 Mks
5. Study of morphological characters of phyla included in theory syllabus. -5 Mks
6. Preparation and study of stratigraphic maps. -3 Mks
7. Sessional. -5 Mks
8. Viva. -5 Mks

B. Sc. IInd year
Practical Syllabus

Laboratory Work

1. Description of Structural models with well labeled diagram.
2. Construction and interpretation of cross section of geological maps.
3. Interpretation of structures with the help of stereographic projections.
4. Morphometric analysis of drainage basins.
5. Description of animal and plant fossils included in the theory syllabus.
6. Locate different stratigraphic formation in the map of India.
7. Megascopic identification of main Igneous, Sedimentary and Metamorphic rocks.
8. Microscoic Identification of main Igneous, Sedimentary and Metamorphic rocks.

B. Sc. IIIrd year
Practical Syllabus

Laboratory Work

1. Physical and optical properties of Ore minerals.
2. Distribution of Economic Ore minerals in the map of India.
3. Megascopic study of Coal and Identification of different rank of coal.
4. Study of Radioactive minerals.
5. Estimation of Ore reserve calculation, Problems of reserve calculation.
6. Study and interpretation of aerial photographs by stereoscope.
7. Study and interpretation of satellite images.

❑❑❑

Other Publications on Geology, GIS and Remote Sensing

S.No.	Title	Author	ISBN	Year
1	A Handbook of Minerals,Crystals,Rocks and Ores	Alexander, P.O.	9788190723787	2009
2	Agrometeorology: A Simplified Textbook	Kathiresan, G.	9789383305568	2015
3	Climate Change and Environment: Concepts and Strategies to Mitigate Impacts	Devesh Sharma & K C Saha	9789385516375	2016
4	Climate Change and Plantations in the Humid Tropics	GSHLV Prasada Rao and C.S. Gopakumar	9789385516368	2016
5	Climate Change and Water Security: Impacts, Future Scenarios, Adaptations and Mitigations	Golam Kibria, A. K. Yousuf Haroon & Dayanthi Nugegoda	9789385516269	2016
6	Computers in Agriculture	Manish K. Sharma & Anil Bhat	9789385516160	2017
7	Disasters: Strengthening Community Mitigation and Preparedness	Khanna, B.K. & Neena Khanna	9789380235455	2011
8	Droughts in Agricultural Production: Monitoring & Management	Rao, G.G.S.N.	9789385516009	2015
9	Encyclopedia of Agricultural Meteorology	Dhaliwal, L.K. & S.S.Hundal	9789380235547	2011
10	Engineering and General Geology	Sawant, P.T.	9789380235516	2011
11	Engineering Hydrology	Pannigrahi, B.	9789385516467	2016
12	Environment,People and Development: Experiences from Desert Ecosystems	Gaur M. Kumar & P.Moharana	9789381450796	2014
13	Environmental Changes and Natural Disasters	Babar, Md.	9788189422752	2007
14	Essentials of Hydrogeology	Gurugnanam, B.	9788190851299	2009
15	Geographic Information System	Gurugnanam, B.	9788190851282	2009
16	Geoinformatics Applications in Agriculture	Singh, Anil Kumar & U.K.Chopra	9788189422233	2007
17	Geological Hazards: Causes,Consequences and Methods of Containments	Ramkumar, Mu.	9788190851275	2009
18	Geomatics in Applied Geomorphology	Ramasamy, SM	9789385516429	2016
19	Geomatics in Tsunami	Ramasamy, SM.: ed.	9788189422318	2006
20	Geospatial Technologies for Natural Resources Management	Soam,S.K. & P.D. Sreekant	9789381450802	2013
21	GIS: Fundamentals,Applications and Implementations	Elangovan, K.	9788189422165	2006
22	Hydrogeomorphology: Fundamentals,Applications and Techniques	Babar, Md.	9788189422011	2005
23	Hyperspectral Remote Sensing and Spectral Signature Applications	Rajendran, S. ed.	9788189422349	2009
24	Mineral Exploration: Recent Strategies	Rajendran, S.: eds.	9788189422714	2007
25	Numerical Methods and Models in Earth Science	Ghosh, Parthasarathi	9789380235417	2011
26	Physical Climatology	Sellers, Williams	9789383305575	2014
27	Practical Agricultural Meteorology	Srivastava, A.K.	9789380235776	2011
28	Practical Isotope Hydrology	Rao, S.M.	9788189422332	2006
29	Precision Farming in Horticulture	Singh, Jitendar, S.K.Jain,L.K.Dashora	9789381450475	2013

30	Principles and Applications of Agricultural Meteorology	Patra, Alok Kumar	9789385516245	2016
31	Remote Sensing Applications in Dryland Natural Resource Management	Gaur, Mahesh	9789381450321	2013
32	Remote Sensing in Geomorphology	Ramasamy, SM: ed.	9788189422059	2005
33	Town Planning Regeneration of Cities	Ashutosh Joshi	9788189422820	2008